Linking the European Union Emissions Trading System

This book focuses on the linking of the European Union Emissions Trading System (EU ETS) with other independent regional ETS.

While rich practical and academic research has evolved on the economic and technical side of ETS linking, political drivers and barriers have so far been underrepresented in this debate. Filling this lacuna and based on international relations theory, existing research, and qualitative fieldwork, this book introduces the range of political conditions that influence linking, such as political leadership and stakeholder activity. Specifically, it analyzes which of these aspects have played a role in three different linking activities of the EU: (1) a failed linking attempt: EU ETS–California Cap-and-Trade Program; (2) a successful linking treaty: EU ETS–Switzerland Emissions Trading System; and (3) an agreed-upon but not realized link: EU ETS–Australia Carbon Pricing Mechanism. Through an interrogation of these examples, Dr. Unger concludes that it is not only the technical challenges or the overall economic benefit but rather domestic interests, structural aspects, and external international political developments that have jointly dominated linking activities, especially those in which the EU takes part.

This book will be of great interest to scholars and policy-makers working in climate policy and EU environmental politics.

Charlotte Unger is a senior research associate at the Institute for Advanced Sustainability Studies, Germany. She acquired her PhD in political science from the Technical University of Munich and has previously worked for various research institutions and think tanks such as the Technical University Berlin, Environmental Action Germany, and the International Carbon Action Partnership.

Routledge Focus on Environment and Sustainability

Sustainable Community Movement Organizations
Solidarity Economies and Rhizomatic Practices
Edited by Francesca Forno and Richard R. Weiner

Climate Change Ethics for an Endangered World
Thom Brooks

The Emerging Global Consensus on Climate Change and Human Mobility
Mostafa M Naser

Traditional Ecological Knowledge in Georgia
A Short History of the Caucasus
Zaal Kikvidze

Traditional Ecological Knowledge and Global Pandemics
Biodiversity and Planetary Health Beyond Covid-19
Ngozi Finette Unuigbe

Climate Diplomacy and Emerging Economies
India as a Case Study
Dhanasree Jayaram

Linking the European Union Emissions Trading System
Political Drivers and Barriers
Charlotte Unger

For more information about this series, please visit: www.routledge.com/Routledge-Focus-on-Environment-and-Sustainability/book-series/RFES

Linking the European Union Emissions Trading System

Political Drivers and Barriers

Charlotte Unger

Routledge
Taylor & Francis Group

LONDON AND NEW YORK

First published 2021
by Routledge
2 Park Square, Milton Park, Abingdon, Oxon OX14 4RN

and by Routledge
52 Vanderbilt Avenue, New York, NY 10017

Routledge is an imprint of the Taylor & Francis Group, an informa business

© 2021 Charlotte Unger

British Library Cataloguing-in-Publication Data
A catalogue record for this book is available from the British Library

Library of Congress Cataloging-in-Publication Data
A catalog record has been requested for this book

ISBN: 978-0-367-42969-0 (hbk)
ISBN: 978-1-003-00043-3 (ebk)

Typeset in Times New Roman
by Newgen Publishing UK

Contents

List of illustrations		vii
Acknowledgments		viii
List of acronyms		ix
1	Introduction	1
2	The fundamentals of linking	15
3	EU climate policy and the design and development of the EU Emissions Trading System: Setting the scene	39
4	The EU Emissions Trading System and the California Cap-and-Trade Program: A failed linking attempt	50
5	A successful linking process between the EU Emissions Trading System and the Switzerland Emissions Trading System	66
6	The EU Emissions Trading System and the Australia Carbon Pricing Mechanism: An agreed-upon, but unrealized link	80
7	Linking is dominated by domestic political interests, domestic structures, and international developments	97

8 Linking in the climate policy debate and future
  prospects 114

  Appendix 126
  Index 128

Illustrations

Figures

1.1 ETS around the world in 2020   2
2.1 The political conditioning factors for linking   23
3.1 Carbon price development in the EU ETS (2008–2019)   41
3.2 Timeline of the EU's linking efforts   45
5.1 Carbon price development in the EU ETS and the SW
ETS (2008–2019)   70

Table

2.1 ETS design elements and their relevance for linking   20

Acknowledgments

My thanks go to Miranda Schreurs (Technische Universität München) and Jørgen Wettestad (Fridtjof Nansen Institute) for devoting their time, expertise, and support to the research process for this book.

I am very grateful for the input, commenting, thought-provoking impulses, and discussions with Adina Spertus-Melhus, Aki Kachi, Alexander Eden, Alexandra Zirkel, Angelika Smuda, Bas Eickhout, Christian Flachsland, Claudia Gibis, Constanze Haug, Dominika Herbst, Frank Jotzo, Jason Gray, Jeff Butler, Jessica Green, Joachim Hein, Julia Hildermeier, Kathleen Mar, Kristian Wilkening, Lars Zetterberg, Lina Li, Marissa Santikarn, Mark Lawrence, Martina Kehrer, Michael Gibbs, Michel Frerk, Peter Zapfel, Sonja Thielges, Tobias Hausotter, William Ackworth, and several anonymous interviewees and reviewers.

I thank Alejandro Monges Zwanck, Elena and Clara Monges Unger, my parents and sisters, as well as many friends and colleagues, for their patience, encouragement, positive energy, and companionship throughout this learning process.

Acronyms

AAU(s)	Assigned Amount Unit(s)
AB 32	Assembly Bill 32 Global Warming Solutions Act (California)
ACCU(s)	Australian Carbon Credit Unit(s)
ADB	Asian Development Bank
AUD	Australian Dollar
AUS CCA	Australian Climate Change Authority
AUS CPM	Australia Carbon Pricing Mechanism
BAFU	Bundesamt für Umwelt (Federal Office for the Environment of Switzerland)
BAU	Business as Usual
BDI	Bund der Industrie (Federation of German Industries)
CAL C&T	California Cap-and-Trade Program
CARB	California Air Resources Board (CARB)
CDM	Clean Development Mechanism
CEFIC	European Chemical Industry Council
CEPI	Confederation of European Paper Industries
CER(s)	Certified Emissions Reduction(s)
CHF	Swiss Francs
COP	Conference of the Parties UNFCCC
CORSIA	Carbon Offsetting and Reduction Scheme for International Aviation
CPLC	Carbon Pricing Leadership Coalition
CPP	Clean Power Plan
DEHSt	Deutsche Emissionshandelsstelle (German Emissions Trading Authority)
DG CL	Directorate General Climate Action
DG ENV	Directorate General Environment
EDF	Environmental Defense Fund
EEA	European Economic Area
EEX	European Energy Exchange

ERU (s)	Emission Reduction Unit(s)
ETS(s)	Emissions Trading, Emissions Trading System(s)
EU	European Union
EUA(s)	European Union Allowance(s)
EU COM	European Commission
EU COU	European Council
EU ETS	European Union Emissions Trading System
EU PA	European Parliament
EUR	Euros
EUROFER	European Steel Association
FDP	Freie Demokratische Partei (Liberal Party Switzerland)
GDP	Gross Domestic Product
GHG	Greenhouse Gas(es)
GOP	US Republican Party, 'Grand Old Party'
ICAO	International Civil Aviation Organization
ICAP	International Carbon Action Partnership
IETA	International Emissions Trading Association
MiTID II	Markets in Financial Instruments Directive II
MoU	Memorandum of Understanding
MRV	Monitoring, Reporting, and Verification
NDC	Nationally Determined Contribution
NDRC	National Development and Reform Commission China
OECD	Organization for Economic Cooperation and Development
QUE C&T	Québec Cap-and-Trade System
RMU(s)	Removal Unit(s)
RGGI	Regional Greenhouse Gas Initiative
SP	Sozialdemokratische Partei der Schweiz (Social Democratic Party Switzerland)
SVP	Schweizer Volkspartei (Party of the People Switzerland)
SW ETS	Switzerland Emissions Trading System
WCI	Western Climate Initiative
UBA	Umweltbundesamt (German Environment Agency)
UNFCCC	United Nations Framework Convention on Climate Change
US EPA	United States Environmental Protection Agency
USD	US Dollar
VAT	Value-Added Tax
WWF	World Wide Fund for Nature

1 Introduction

The 21st century has thus far been characterized by a curvy road of progress in international cooperation on action to address climate change. Yet at the same time, it has also emerged as a stage for experimentation with new forms of climate governance, including close regional cooperation. Among the many available climate policy tools, the pricing of greenhouse gas emissions (GHG), for example through emissions trading systems (ETSs),[1] has become the instrument of choice for many regions in the world. An ETS is a market-based instrument that has the overall aim of reducing GHG through establishing a price for carbon dioxide (CO_2) or carbon dioxide equivalents (CO_2e). Participating companies have to hold an allowance or credit for every ton of emitted emissions, which is generally obtained for free from the government or bought via auctioning. An overall cap guarantees a fixed level of emissions and ideally results in an emission reduction pathway. Today, 21 independent ETSs are in place (ICAP, 2020a) (see Figure 1.1). As of 2019, carbon pricing initiatives worldwide covered around 11 GtCO2e, representing 20.1% of global GHG emissions (The World Bank, 2019).

While these instruments feature unique designs, they all set a priced legal entitlement for each ton of GHG, which can be traded on a domestic carbon market. Some regions go one step further and create common markets by 'linking' their ETSs. *Linking* is a bi- or multilateral cooperation, where ETS-operating jurisdictions create a common carbon market. This allows covered entities to comply with their domestic climate policy obligations by using the GHG entitlements from the other system or systems. In this book, linking refers to the entire activity and process that two or more jurisdictions undertake to connect their ETSs, starting with informal bilateral talks and ending with the ratification of a linking agreement.

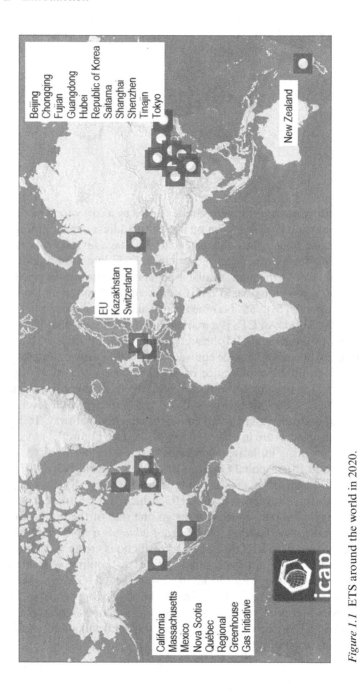

Figure 1.1 ETS around the world in 2020.

Source: ICAP (2020): ETS Map. Retrieved 28.07.2020 from https://icapcarbonaction.com/en/ets-map. Updated 23.06.2020 (used with permission granted by the ICAP Secretariat)

ETS linking – an economic opportunity for climate policy?

In theory, linking can increase the efficiency of ETSs and could be an option for improving international climate policy. The economic rationale for linking has dominated the political and academic debate. In the event of linking, a larger common carbon market is created and companies can choose from a larger pool of abatement opportunities. The crucial benefit of ETS as a climate policy instrument—the reduction of emissions where it is most cost efficient—could be strengthened through linking. Linking has the effect of leveling the marginal emissions abatement costs of the linking partners and bringing about convergence in the carbon price (Flachsland et al., 2009a, Green et al., 2014a, Haites, 2014). Overall, ETS linking should produce cost savings for the linked markets, as the collective goal of a combined emissions cap could be achieved more cost efficiently (Haites and Mullins, 2001, Burtraw et al., 2013, Kachi et al., 2015, Ellis and Tirpak, 2006, Tuerk et al., 2009, Jaffe et al., 2009). Thus, in principle more ambitious climate mitigation could be achieved at the same costs and jurisdictions could advance their efforts to mitigate climate change (Green et al., 2014a, Mehling et al., 2018). Linking could boost climate-political ambition in general; for example, if countries strengthened their emissions reduction targets or added new mitigation tools to their policy mix after linking (Santikarn et al., 2018, Mehling et al., 2018). Additionally, it is argued that a linked carbon market generally functions better, because it can improve important aspects such as market liquidity, price volatility, and level competitive distortions and the influence of large market participants (Flachsland et al., 2009b, Jaffe and Stavins, 2008, Flachsland et al., 2009a, Edenhofer et al., 2007, Ellis and Tirpak, 2006, Haites, 2014).

Additionally, the value of linking can also be seen in its political and symbolic meaning. International climate policy can in theory benefit from linking, because it facilitates cooperation and learning between jurisdictions and may incentivize other countries to follow the example of ETS climate policy (Santikarn et al., 2018, Bodansky et al., 2016). The process of linking and the subsequent knowledge exchange helps foster the distribution of good practices. It can incentivize other countries to implement ETSs and ultimately achieve a 'globally connected' carbon market that is desirable for countries to be part of. Moreover, international linking agreements signal to the world, but also to domestic constituents, a strong commitment to the common fight against climate change (Flachsland et al., 2009a, Green et al., 2014b, Bodansky et al., 2016).

In spite of this rationale for linking, it is important to consider that the generally positive view of linking has more recently been challenged by several academic scholars, such as Green (2017) and Gulbrandsen et al. (2019). These critiques argue that the costs of linking have been underestimated and that linking does not necessarily have a positive effect for all partners and overall climate policy. Rather, they argue, the political reality has led to regionally heterogeneous and very individualized ETS approaches that obstruct linking (Gulbrandsen et al., 2019, Green, 2017). The understanding of this academic dispute is part of the objective of this book. The book therefore also examines practical cases where linking was seen as disadvantageous and considers the difficulties and risks named above.

Recent policy developments provide an impetus for research on linking

When looking at the evolving landscape for carbon pricing instruments, a few observations can be made that build the case for further analysis of linking.

Cooperative efforts to combat climate change have been progressing only very slowly. The difficulties involved in negotiating a worldwide agreement under the United Nations Framework Convention on Climate Change (UNFCCC) also demonstrate that consensus on the need for a globally coordinated approach is still lacking. Most countries still prefer to pursue national or regional approaches over committing themselves to an international treaty. This tendency is also reflected in the Paris Agreement of 2015. Here, instead of a central distribution of the necessary responsibilities for reducing emissions to achieve the 'well below' 2°C objective between all parties, each country individually submitted the mitigation it considered feasible as its Nationally Determined Contribution (NDC).[2] Shortly after the adoption of the Paris Agreement, it became clear that these climate change mitigation promises were far from sufficient; even if all NDCs were successfully implemented, they would only achieve one-third of the minimum mitigation to keep global warming at 2°C (UN Environment, 2018).

In light of the difficulties in agreeing on international cooperation through a global treaty, many experts have debated whether linking different market-based policy instruments, like ETSs, could offer a way out of the gridlock and provide a new bottom-up architecture for international climate policy (Ranson and Stavins, 2012, Jaffe et al., 2009, Comendant and Taschini, 2014). Research work thus has a key role to play in examining whether and how linking of ETSs can support the progress of international climate policy, for example, through providing additional political structures.

Several global developments hint at a political interest in linking. While the Paris Agreement's text does not explicitly mention the linking of ETSs or other carbon pricing mechanisms, it does contain provisions that leave room for this possibility. The provisions in Article 6 enable the transfer of emissions reduction units, and therefore offer the potential to enhance bilateral and multilateral forms of cooperation on carbon pricing (EDF and IETA, 2016). This idea is not new. The Kyoto Protocol had previously established international emissions trading and mechanisms, such as Clean Development Mechanism (CDM), to allow for the transfer of emissions reductions between its parties. In the Paris Agreement, international reduction credits can be used to fulfill each country's NDC, independent of whether these units come from a different ETS, a crediting mechanism (for example, a 'new CDM') or any other measured and verified emission reduction unit. While the EU to date has not publicly considered such provisions, several other countries plan to accept the use of international reduction credits, such as New Zealand or Switzerland.

Independent of these international frameworks, many countries, regions or subnational entities that operate an ETS, or are considering its introduction, have expressed an interest in linking and some have initiated links. For example, in North America, the Regional Greenhouse Gas Initiative (RGGI) connects different US states in their efforts to regulate GHG from the electricity sector. Some South American jurisdictions have also shown interest in linking. For instance, in October 2015, Panama announced that it was considering the introduction of an ETS and was interested in linking with other carbon pricing initiatives of the region (Carbon Pulse, 2015a). Mexico and Québec signed a cooperation agreement in 2015, also exploring a plan to link the Mexican carbon market through the provision of offsets to California and Québec (Carbon Pulse, 2015b). In the EU, in addition to negotiations over the linking of the EU Emissions Trading System (EU ETS) to Switzerland's ETS (SW ETS), policy-makers have begun discussing a potential link to a Chinese national ETS (Cuff, 2016). Interest in linking is also visible in other parts of Asia and Oceania, with the ETS-implementing countries of Kazakhstan, South Korea and New Zealand all advocating linking policy (Carbon Pulse, 2016).

In addition, other activities by national governments, as well as transnational actors such as the World Bank, the Asian Development Bank (ADB) and the Organization for Economic Cooperation and Development (OECD), demonstrate a growing interest in the topic.

These can take the form of commissioned studies, international conferences and meetings, and financing for capacity building and training activities related to the topic. Various international networks that deal with linking have also been launched, such as the International Carbon Action Partnership (ICAP), the Carbon Pricing Leadership Coalition (CPLC) and the World Bank Group's Networked Carbon Markets initiative.

Academic research builds the case for linking but largely overlooks its political side

Ever since the launch of the first ETSs, such as the sulfur dioxide reduction component of the US Acid Rain Program, academic literature in this field has prospered and is today quite extensive. Academics have focused not only on research dealing with the functioning of market-based mechanisms and ETSs, but also examinations of their impact, effectiveness, and underlying political processes. Research interest in linking has experienced highs and lows but persisted over time. One branch of literature addresses *linking within international climate policy*, for example through international market mechanisms, such as international emissions trading and the CDM (Michaelowa, 2011). Several papers also discuss linking in the context of the international climate policy negotiations and its potential for international bottom-up cooperation (Ranson and Stavins, 2012, Victor, 2015, Bodansky et al., 2016, Green et al., 2014a). Following the Paris Agreement, a new academic discussion has also emerged, envisioning what linking under Article 6 of the Agreement could look like (Mehling et al., 2018).

A voluminous body of literature also examines the *linking of independent domestic ETSs*. The majority of these texts deal with the compatibility of different ETS designs. Most aim at revealing what happens when two ETSs that differ in their design link, e.g. if they have different offset regulations or cover different sectors (Kachi et al., 2015, Hawkins and Jegou, 2014, Burtraw et al., 2013, Zetterberg, 2012, Anger, 2008, Tuerk et al., 2009, Edenhofer et al., 2007, Ellis and Tirpak, 2006, Baron and Bygrave, 2002, Haites and Mullins, 2001, Sterk et al., 2006a, Gulbrandsen et al., 2019). They examine which issues arise as a result of these differences and which design features should be harmonized prior to a linkage. Often, ETS designs are discussed under specific s*cenarios of linked carbon markets*. Many studies draw scenarios of prospectively linked carbon markets in different regions worldwide (Hawkins and Jegou, 2014, Zetterberg, 2012, Tuerk et al., 2009, Flachsland et al., 2009b, Goers et al., 2012, Haites and Mehling, 2009a, Blyth and Bosi,

2004). In addition to the extensive discussion on the consequences and constraints of linking, technical solutions that seek to smooth some of these unwanted effects have also been proposed; this body of literature presents *options for restricted linking*, e.g. exchange quotas for emissions allowances, price rates, linking by degrees or restricted linking (Schneider et al., 2017, Burtraw et al., 2013). Ultimately, the majority of studies on linking express a rather optimistic view on linking; there is a common understanding that linking is at least technically feasible.

Notwithstanding the rich technical and economic debate on linking, many authors overlook the difficulties that arise related to political practice and governance of linking. Only very few texts concentrate on political barriers for linking, such as differing policy objectives (Gulbrandsen et al., 2019, Flachsland et al., 2009a, Green et al., 2014a). Several studies focus on the political processes that led to the implementation of ETSs (Christiansen and Wettestad, 2003, Skjærseth and Wettestad, 2008, Orlowski and Gründinger, 2011, Convery, 2009). More recently, several reports and discussion papers have also been published on the governance process of linking (Santikarn et al., 2018, Tänzler et al., 2018, Görlach et al., 2015, Beuermann et al., 2017). Many of the above cited papers either mention that ETS differences, such as the carbon price level, lead to political difficulties, or they refer to ETS design-related political sensitivities that arise with linking. However, only very few use the political facet of linking as a starting point or conduct a substantial analysis of linking's political constraints and drivers. The political conditions under which ETSs exist (such as, for example: *the political system, stakeholders*, and their objectives that may push or limit linking, *the economic motivation* behind setting a certain carbon price, *the legal framework, international developments*) are tackled only on a very superficial level. Among the political constraints identified in the literature to be essential for understanding linking are, for instance, differing policy targets and ambition, distribution of financial gains and losses, the loss of control and autonomy, legal and systemic structures, geographical proximity and stakeholder opposition to the system (Green et al., 2014a, Ranson and Stavins, 2016, Gulbrandsen et al., 2019, Wettestad and Jevnaker, 2014, Flachsland et al., 2009a, Müller and Slominski, 2016, Green, 2017).

Academic literature on linking lacks an approach to identifying the whole range of political variables that influence linking. Furthermore, research on linking does not systematically examine the political conditions that have been at work in past linking negotiations or linking attempts. However, an in-depth analysis of such case studies, one that examines why and when certain factors influenced the decisions of policy-makers to link or not, is needed. Such an analysis could help paint a much more holistic picture of linking, complementing existing knowledge on linking's technical and economic dimensions.

Linking lacks progress in practice

The above paragraphs have highlighted the benefits of linking, as well as the climate-political momentum behind it. Additionally, theoretical arguments state that even despite the technical and economic challenges, linking could be beneficial. Given all this, one would expect a boom in linking. However, the fact is that, until recently, very few examples of fully realized linkages between independent ETSs can be found. Two notable exceptions are the 2014 linkage between the California Cap-and-Trade Program (CAL C&T) and the Québec Cap-and-Trade System (QUE C&T), and the very recent link between the EU and Switzerland in 2020. Despite the existence of 21 operating ETSs worldwide as of 2020, concrete plans for bi-or multilateral linkages are rare. Although the playing field has continued to evolve, existing systems have matured, and interest in linking has been communicated on many occasions, many of the most established systems have been hesitant to link; no direct links exist between the EU ETS, RGGI, CAL C&T, South Korea ETS, or the New Zealand ETS.

This situation raises the empirical questions motivating this book: *Why do jurisdictions sometimes link their ETSs and sometimes not?* And furthermore, why do we find such discrepancies between academic research and political practice? Why have jurisdictions, despite the potential benefits and (theoretical) technical feasibility of linking, not been more active? Why has interest in linking in some cases only led to informal talks, while in others it has resulted in informal agreements or the conclusion of a linking treaty?

About this book's approach

"Linking is a multi-faceted policy decision (…)" (Ranson and Stavins, 2016: 286). This statement can be taken as starting point for this book, which seeks to explain linking from an international relations and political science perspective. To answer the questions raised above, it focuses on the linking activities of the EU. The central argument of this book is that, in addition to linking's well-known technical and economic challenges, domestic politics are inextricably interrelated and are, at present, the deciding factor in whether a particular link will be realized. This book explains how political structures, political priorities and interests, as well as international developments—such as *economic interests, leadership, power and ETS size, agenda setting, stakeholder activity* and other variables—have influenced recent linking processes in the EU and led to their current state and outcomes: an agreement or not. This analysis follows rational choice model conceptualizations, under which jurisdiction will only link if they regard the cooperation as beneficial overall.

At the center of the book's analysis stand three case studies involving the EU ETS and other independent regional ETSs between 2008 and 2019: first, a potential although unrealized linkage between the EU and California (EU ETS–CAL C&T: 2008–2011); second, a successful case, with an operating link between the EU and Switzerland (EU ETS–SW ETS: 2008–2019); and finally a case in which linking was agreed on but never realized, between the EU and Australia (EU ETS–Australia Carbon Pricing Mechanism (AUS CPM): 2011–2014). For each case, the political conditions are examined in order to explain the success or failure of the linking attempt.

Methodologically, the research carried out for this book followed an inductive and explorative research structure that aims to systematize the political conditions for linking and thereby build generalizable hypotheses for future analyses. Data was extracted from previous academic research as well as nonacademic material data such as reports, conference protocols and press releases. Additionally, 19 semistructured interviews with academic and political experts from the examined jurisdictions were used to inform the analytical framework and case studies of this book (see Appendix for further details).

This book employs a comparative perspective between its three case studies and can be classified as a meta-study. It provides detailed case analyses, while at the same time seeking out commonalities among the cases that build up to generally applicable patterns. The primary aim of this book is to make an empiric contribution: concretely addressing the policy debate on carbon markets in order to shed light on what conditions were at play in past linking activities and which factors may thus be prerequisites for linking in practice. In doing so, this approach also contributes to filling a gap in the academic and conceptual debate on linking through systematizing and expanding the academic discussion with knowledge from political science. It complements knowledge on the political dimension of linking by proposing a framework that can be adapted for future research and further case studies.

Furthermore, this book contributes and connects to several broader discussions. It engages with recent political debates on the UNFCCC process and Paris Agreement that explore how and what forms of cooperation can be realized in order to strengthen existing mitigation commitments. This contributes to the current debate on the state of the global climate governance landscape, for example by providing insights on the carbon market politics of certain crucial jurisdictions.

Last but not least, this book draws from one of the oldest fields in the international relations: cooperation and under what conditions it occurs. For this discussion, this work not only provides empirical 'food for thought' on the necessary conditions for cooperation but offers the case studies as an impetus for future research on how international cooperation between multilevel governance actors works.

Following this introduction, the second chapter of this book explains how linking works and which forms linking activities can take in practice. Here, building on international relations and international climate policy theory, a framework that compiles the *conditioning factors* for linking is conceptualized. The third chapter introduces the EU ETS; explores its development, design, and the role of relevant institutions; and notes the EU's first linking attempts. Following on this, three empirical chapters examine the three case studies, namely the EU's linking initiatives with California, Australia, and Switzerland. Each case study contains a description of the development and design of the respective ETS and then examines the conditioning factors that influenced the linking activity with the EU ETS. The penultimate chapter provides a comparative perspective on the three case studies and presents the conclusive findings on why EU linking activities have reached their current states, and why global linking progress continues to lag behind. These assumptions are contextualized within the broader climate policy debate in the last chapter of this book. It considers how linking might experience an upturn in the mid- to long-term future, revisiting another potential ETS partner constellation with China, and alternative linking approaches, such as restricted linking.

Notes

1 Two different terms are almost synonymously used for the same instrument type: Emissions trading system and cap-and-trade system. While the first simply refers to a system that covers emissions, the second specifies that the instrument includes a certain limit or 'cap' in addition to the trade that occurs. Almost all operating ETSs include all of these characteristics. For simplification, this book will use the abbreviation 'ETS' for both terms, emissions trading system/s and cap-and-trade system/s.
2 The Kyoto Protocol centrally distributed emissions budgets and the necessary responsibilities for reducing emissions to industrialized countries.

References

ANGER, N. 2008. Emissions trading beyond Europe: Linking schemes in a post-Kyoto world. *Energy Economics*, 30, 2028–2049.

BARON, R. & BYGRAVE, S. 2002. *Towards international emissions trading: Design implications for linkages* [Online]. Available: www.oecd.org/environment/cc/2766158.pdf [Accessed 17.04.2019].

BEUERMANN, C., BINGLER, J., SANTIKARN, M., TÄNZLER, D. & THEMA, J. 2017. *Considering the effects of linking emissions trading schemes* [Online]. Berlin: German Emissions Trading Authority (DEHSt). Available: www.dehst.de/SharedDocs/downloads/EN/emissions-trading/Linking_manual.pdf?__blob=publicationFile&v=3 [Accessed 20.03.2019].

BLYTH, W. & BOSI, M. 2004. *Linking non-EU domestic emissions trading schemes with the EU emissions trading scheme* [Online]. IEA, OECD Information Paper. Available: www.iea.org/textbase/papers/2004/non_eu.pdf [Accessed 13.12.2014].

BODANSKY, D. M., HOEDL, S. A., METCALF, G. E. & STAVINS, R. N. 2016. Facilitating linkage of climate policies through the Paris outcome. *Climate Policy*, 16, 956–972.

BURTRAW, D., PALMER, K. L., MUNNINGS, C., WEBER, P. & WOERMAN, M. 2013. *Linking by degrees: Incremental alignment of cap-and-trade markets* [Online]. Resources for the Future 13-04. Available: www.rff.org/publications/working-papers/linking-by-degrees-incremental-alignment-of-cap-and-trade-markets/ [Accessed 12.03.2017].

CARBON PULSE. 2015a. *Panama considering carbon trading amid rapid growth, shipping concerns* [Online]. Available: https://carbon-pulse.com/11176/ [Accessed 20.7.2016].

CARBON PULSE. 2015b. *Mexico, Quebec to cooperate on carbon markets* [Online]. Available: http://carbon-pulse.com/6143/ [Accessed 11.07.2016].

CARBON PULSE. 2016. *BRIEFING: The challenges of linking Asia-Pacific carbon markets: Asia Pacific.* [Online]. Available: http://carbon-pulse.com/18099/ [Accessed 28.08.2016].

CHRISTIANSEN, A. C. & WETTESTAD, J. 2003. The EU as a frontrunner on greenhouse gas emissions trading: How did it happen and will the EU succeed? *Climate Policy*, 3, 3–18.

COMENDANT, C. & TASCHINI, L. 2014. *Submission to the inquiry by the House of Commons Select Committee on Energy and Climate Change on 'Linking Emissions Trading Systems'* [Online]. Centre for Climate Change Economics and Policy Grantham Research Institute on Climate Change and the Environment. Available: www.cccep.ac.uk/Publications/Policy/docs/Comendant-and-Taschini-policy-paper-April-2014.pdf [Accessed 20.07.2016].

CONVERY, F. J. 2009. Origins and development of the EU ETS. *Environmental Resource Economics*, 43, 391–412.

CUFF, M. 2016. What would a link-up between China and the EU's carbon markets mean for emissions trading? Available from: www.businessgreen.

com/bg/analysis/2443229/what-would-a-link-up-between-china-and-the-eu-s-carbon-markets-mean-for-emissions-trading   [Accessed   15.08.2016].

EDENHOFER,   O.,   FLACHSLAND,   C.   &   MARSCHINSKI,   R. 2007. *Towards a global CO$_2$ market* [Online]. Available: https://pdfs. semanticscholar.org/29fd/f66e2341bc63d2d57b9b431a42ddf2a328b5.pdf [Accessed 03.02.2020].

EDF & IETA. 2016. *Carbon pricing: The Paris Agreement's key ingredient* [Online]. Available: www.edf.org/media/governments-can-surpass-paris-climate-pledges-through-markets [Accessed 14.04.2017].

ELLIS, J. & TIRPAK, D. 2006. *Linking GHG emission trading systems and markets* [Online]. Organisation for Economic Co-operation Development, International Energy Agency. Available: www.oecd.org/environment/cc/ 37672298.pdf [Accessed 03.03.2019].

FLACHSLAND, C., MARSCHINSKI, R. & EDENHOFER, O. 2009a. To link or not to link: Benefits and disadvantages of linking cap-and-trade systems. *Climate Policy*, 9, 358–372.

FLACHSLAND, C., MARSCHINSKI, R. & EDENHOFER, O. 2009b. Global trading versus linking: Architectures for international emissions trading. *Energy Policy*, 37, 1637–1647.

GOERS, S. R., PFLÜGLMAYER, B. & LUGER, M. J. 2012. Design issues for linking emissions trading schemes—a qualitative analysis for schemes from Europe, Asia and North America. *Journal of Environmental Science and Engineering*, 1, 1322.

GÖRLACH, B., MEHLING, M. & ROBERTS, E. 2015. *Designing institutions, structures and mechanisms to facilitate the linking of emissions trading schemes* [Online]. Berlin: German Emissions Trading Authority (DEHSt). Available: www.dehst.de/SharedDocs/downloads/EN/perspectives/Linking_report.pdf?__blob=publicationFile&v=3 [Accessed 02.03.2020].

GREEN, J. F. 2017. Don't link carbon markets. *Nature News*, 543, 484.

GREEN, J. F., STERNER, T. & WAGNER, G. 2014a. A balance of bottom-up and top-down in linking climate policies. *Nature Climate Change*, 4, 1064–1067.

GREEN, J. F., STERNER, T. & WAGNER, G. 2014b. *The politics of market linkage: Linking domestic climate policies with international political economy* [Online]. Fondazione Eni Enrico Mattei Nota di Lavoro 64.2014. Available: www.econstor.eu/bitstream/10419/102003/1/NDL2014-064.pdf [Accessed 03.04.2018].

GULBRANDSEN, L. H., WETTESTAD, J., VICTOR, D. G. & UNDERDAL, A. 2019. The political roots of divergence in carbon market design: Implications for linking. *Climate Policy*, 19, 427–438.

HAITES, E. 2014. *Lessons learned from linking emissions trading systems: General principles and applications* [Online]. Washington DC: Partnership for Market Readiness (PMR). Available: www.thepmr.org/system/files/ documents/PMR%20Technical%20Note%207.pdf [Accessed 09.09.2019].

HAITES, E. & MEHLING, M. 2009a. Linking existing and proposed GHG emissions trading schemes in North America. *Climate Policy*, 9, 373–388.

HAITES, E. & MULLINS, F. 2001. *Linking domestic and industry greenhouse gas emission trading systems* [Online]. Margaree Consultants. Available: http://citeseerx.ist.psu.edu/viewdoc/download?doi=10.1.1.512.9323&rep=re p1&type=pdf [Accessed 12.12.2018].

HAWKINS, S. & JEGOU, I. 2014. *Linking emissions trading schemes: Considerations and recommendations for a joint EU-Korean carbon market* [Online]. ICTSD Global Platform on Climate Change, Trade and Sustainable Energy Available: www.ictsd.org/sites/default/files/research/ linking-emissions-trading-schemes-considerations-and-recommendations-for-a-joint-eu-korean-carbon-market.pdf [Accessed 09.09.2019].

ICAP. 2020a. *Emissions trading worldwide: International Carbon Action Partnership (ICAP) status report 2019* [Online]. Berlin. Available: https:// icapcarbonaction.com/en/?option=com_attach&task=download&id=677 [Accessed 02.05.2020].

JAFFE, J., RANSON, M. & STAVINS, R. N. 2009. Linking tradable permit systems: A key element of emerging international climate policy architecture. *Ecology LQ,* 36, 789.

JAFFE, J. & STAVINS, R. N. 2008. *Linkage of tradable permit systems in international climate policy architecture* [Online]. National Bureau of Economic Research. Available: www.nber.org/papers/w14432.pdf [Accessed 02.05.2020].

KACHI, A., UNGER, C., BÖHM, N., STELMAKH, K., HAUG, C. & FRERK, M. 2015. *Linking emissions trading systems: A summary of current research* [Online]. International Carbon Action Partnership Policy Paper. Available: https://icapcarbonaction.com/en/?option=com_ attach&task=download&id=575 [Accessed 02.05.2020].

MEHLING, M. A., METCALF, G. E. & STAVINS, R. N. 2018. Linking climate policies to advance global mitigation. *Science,* 359, 997–998.

MICHAELOWA, A. 2011. Failures of global carbon markets and CDM? *Climate Policy,* 11:1, 839–841. DOI: 10.3763/cpol.2010.0688

MÜLLER, P. & SLOMINSKI, P. 2016. Theorizing third country agency in EU rule transfer: Linking the EU Emission Trading System with Norway, Switzerland and Australia. *Journal of European Public Policy,* 23, 814–832.

ORLOWSKI, M. & GRÜNDINGER, W. 2011. Der Streit um heiße Luft: Der Einfluss von Interessengruppen auf den EU-Emissionshandel und seine Umsetzung in Deutschland und dem Vereinigten Königreich. *Zeitschrift für Public Policy, Recht und Management,* 4, 125–148.

RANSON, M. & STAVINS, R. N. 2012. *Post-Durban climate policy architecture based on linkage of cap-and-trade systems* [Online]. National Bureau of Economic Research. Available: www.belfercenter.org/sites/default/files/files/ publication/ranson-stavins_dp51.pdf [Accessed 13.03.2020].

RANSON, M. & STAVINS, R. N. 2016. Linkage of greenhouse gas emissions trading systems: Learning from experience. *Climate Policy,* 16, 284–300.

SANTIKARN, M., LI, L., THEUER, S. L. H. & HAUG, C. 2018. *A guide to linking emissions trading systems* [Online]. Berlin: Report published by ICAP. Available: www.icapcarbonaction.org/publications [Accessed 03.09.2019].

SCHNEIDER, L., LAZARUS, M., LEE, C. & VAN ASSELT, H. 2017. Restricted linking of emissions trading systems: Options, benefits, and challenges. *International Environmental Agreements: Politics, Law and Economics*, 17, 883–898.

SKJÆRSETH, J. B. & WETTESTAD, J. 2008. Implementing EU emissions trading: Success or failure? *International Environmental Agreements: Politics, Law, Economics*, 8, 275–290.

STERK, W., BRAUN, M., HAUG, C., KORYTAROVA, K. & SCHOLTEN, A. 2006a. *Ready to link up?: Implications of design differences for linking domestic emissions trading schemes* [Online]. Available: https://epub.wupperinst.org/frontdoor/index/index/docId/2495 [Accessed 02.05.2020].

TÄNZLER, D., SANTIKARN, M., STELMAKH, K., KACHI, A., BEUERMANN, C., THEMA, J., HAUPTSTOCK, D. & BINGLER, J. 2018. *Analysis of risks and opportunities of linking emissions trading systems* [Online]. German Environment Agency (UBA). Climate Change 07/2018. Available: www.umweltbundesamt.de/sites/default/files/medien/1410/publikationen/2018-02-23_climate-change_07-2018_linking-eu-ets.pdf [Accessed 02.03.2019].

THE WORLD BANK. 2019. *Carbon pricing dashboard* [Online]. Available: https://carbonpricingdashboard.worldbank.org/ [Accessed 05.05.2019].

TUERK, A., MEHLING, M., FLACHSLAND, C. & STERK, W. 2009. Linking carbon markets: Concepts, case studies and pathways. *Climate Policy*, 9, 341–357.

UN ENVIRONMENT. 2018. *Emissions gap report 2018* [Online]. United Nation Environment Programme. Available: http://wedocs.unep.org/bitstream/handle/20.500.11822/26895/EGR2018_FullReport_EN.pdf?sequence=1&isAllowed=y [Accessed 04.04.2019].

VICTOR, D. G. 2015. *The case for climate clubs* [Online]. International Centre for Trade and Sustainable Development (ICTSD) Available: www.e15initiative.org/ [Accessed 27.07.2018].

WETTESTAD, J. & JEVNAKER, T. 2014. The EU's quest for linked carbon markets. *In:* CHERRY, T. L., HOVI, J. & MCEVOY, D. M. (eds.) *Toward a new climate agreement: Conflict, resolution and governance.* London: Routledge.

ZETTERBERG, L. 2012. *Linking the emissions trading systems in EU and California* [Online]. Fores, Swedish Environmental Research Institute. Available: https://fores.se/wp-content/uploads/2013/04/FORES-California_ETS-web.pdf [Accessed 05.05.2019].

2 The fundamentals of linking

What is linking? The linking process and formats of linking

Linking refers to the direct connection and interaction between carbon regulation policies. In most cases, the term linking is used to describe when two or more ETSs officially agree on a connection between each other that allows them to exchange emission allowances. Participants of one of the ETSs can use allowances issued by the other scheme or schemes to meet their obligations under their home scheme. The linking systems create a common, larger carbon market.

Systems can either link directly or indirectly. Two forms of direct links should be distinguished: *bilateral/multilateral* or *unilateral linking*. When two ETSs link bilaterally, either system may directly use allowances issued by the other ETS for compliance. If more than two systems are involved, this link is considered multilateral. However, if only one scheme establishes a link accepting external credits for compliance, it is referred to as unilateral linking. A bilateral link can also comprise two jurisdictions establishing mutual unilateral links with one another (Mehling and Görlach, 2016, Burtraw et al., 2013). Unilateral linking can be realized through the acceptance of either credits from another ETS or the certificates of an offset or crediting mechanism. As of present, the latter is the most common approach for unilateral linking, e.g. the EU ETS's acceptance of Certified Emissions Reductions (CER) credits originating from the CDM.

Indirect linking happens, if two systems are connected through a third party. Examples include the common acceptance of an offset standard, e.g. the CDM, or when one of the linked ETSs is also linked to a third system. The 'third party' or mechanism will therefore still have an impact on the scheme that does not directly accept its credits (Hawkins and Jegou, 2014). Another distinction can be made regarding to the degree to which certificates can be exchanged. The term 'full linking' usually refers to systems that link bilaterally and accept credits of the linking partner directly and without restriction. In contrast, systems can also agree to only link parts of their systems, for instance certain sectors, or define an amount of exchange that can take place. This is referred to as 'restricted linking'. For example, this could be realized through limiting the amount of exchanged credits (Burtraw et al., 2013) or through imposing an exchange rate for allowances (Comendant and Taschini, 2014).

This book focuses on direct bilateral linking: in other words, a connection in which all ETSs may exchange emissions permits without restriction, excluding unilateral and indirect linking. *Linking* refers to an activity or process, whereas a *link*, *linkage*, or *linking agreement* refers to the desired outcome as part of this process. Several authors, such as Santikarn et al. (2018), have sought to identify different phases of the linking process. This book divides the process of linking into four phases: (1) *informal talking and negotiating phase*, in which jurisdictions start to talk about a connection between their ETSs and proceed to carry out a deeper knowledge exchange, often making this interest publicly known, e.g. through a public announcement or press release; (2) *formal interjurisdictional negotiations phase*, where official bilateral negotiations start, often requiring a formal mandate, and where the focus lies on agreeing on the technical details of the link; (3) *agreement phase*, where a common understanding among all parties is reached and a linking agreement is drafted and adopted, e.g. through the ratification of a treaty; and (4) *maintenance and operation phase*, which begins when the domestic legislation is adapted to permit the accounting of the partner's allowances and the operation of the common market. These four phases take place on multiple governmental levels and entail a large variety of institutions and stakeholders.

The legal structures of linking

In essence, linking can take two broad legal structures: agreements based on international law or agreements that depend on reciprocal rules in the domestic legislation. The agreements based on international law are binding international treaties. An international treaty has the advantage of being legally binding. Furthermore, it provides a high degree of certainty and reliability to domestic and international politics, because it cannot be amended or terminated unilaterally, threatening sanctions in case of default (Lenz et al., 2014, Mehling, 2016, Haites, 2014). However, international agreements may be difficult or even impossible to achieve, because not all ETS operating regions, e.g. subnational entities such as the US states, or cities, are subjects to international law (Haites and Mehling, 2009, Lenz et al., 2014, Mehling, 2016).

The alternative to an international treaty is a nonbinding arrangement of political nature. In order to achieve such arrangements, the linking jurisdictions could issue joint statements or create a Memorandum of Understanding (MoU) (Mehling, 2016, Mehling and Haites, 2009, Lenz et al., 2014, Görlach et al., 2015). Generally, such political decisions are easier to reach, because they don't require ratification and are easy to alter or to terminate. Still, to guarantee the functioning of a linked carbon market and to provide more predictability for market participants, some sort of binding regulations will probably necessary. Otherwise, the link could be altered too easily and without the consent of the linking partner, for instance through a government change (Mehling, 2016, Mehling and Haites, 2009, Lenz et al., 2014, Görlach et al., 2015). Therefore, if the bi- or multilateral arrangement is only of political nature, provisions in the domestic legislation are necessary in order to make a link more stable (Mehling, 2016, Mehling and Haites, 2009, Lenz et al., 2014, Görlach et al., 2015).

A glimpse into research that focuses on the technical design compatibility of ETSs

Even before the launch of the EU ETS academic research has started to examine linking from a technical perspective. These studies aim at revealing what happens when two ETSs that differ in their design link (Kachi et al., 2015, Hawkins and Jegou, 2014, Burtraw et al., 2013, Zetterberg, 2012, Anger, 2008, Tuerk et al., 2009, Edenhofer et al., 2007, Ellis and Tirpak, 2006, Baron and Bygrave, 2002, Haites and Mullins, 2001, Sterk et al., 2006a, Gulbrandsen et al., 2019). Many studies draw scenarios and build case studies for potential links (Hawkins and Jegou, 2014, Goers et al., 2012, Haites and Mehling, 2009, Sterk et al., 2009, Tuerk et al., 2009, Zetterberg, 2012, Flachsland et al., 2009b, Burtraw et al., 2013, Blyth and Bosi, 2004, Sterk et al., 2006a). Specifically, this research identifies which issues arise and which design features should be harmonized prior to a linkage. Based on this body of literature, essential design features of ETSs have been recognized to be:

• The *carbon or allowance price* is created when the government sells allowances through auctions (on the primary market) or at the moment companies trade allowances on the secondary market. After linking, the carbon prices of the partner ETSs converge and all participating companies face the same carbon price (Tuerk et al., 2009).

- The *emissions cap* is the limit or total amount of GHG emissions allowed to be emitted by the entities covered by the ETS over a certain period of time. This overall emissions budget is determined by the jurisdiction's ambition for emissions reductions and its policy goals (ICAP, 2015b). Often, a *cap reduction factor* ensures that the cap declines over time and defines the emission reduction pathway for the ETS. The caps of the linking partners can be combined in a linked market. The level or ambition of the cap contributes to the ETS' environmental stringency. Linking systems with different levels of stringency, could compromise the environmental effectiveness of the linked market, for example, in the case of a link between two systems with an absolute and intensity-based cap, respectively (Haites and Mullins, 2001, Green et al., 2014a, Edenhofer et al., 2007, Sterk and Schüle, 2009, Tuerk et al., 2009, Haites, 2014).

- The *method used for of allocating of allowances*: In an ETS, allowances are distributed to the covered entities either for free or they are sold by the government. Free allocation commonly is based on grandfathering or output based allocation (Santikarn et al., 2018). Usually, a government releases allowances for purchase through auction. From a technical perspective, the allocation methodology is not a challenge to linking (Tuerk et al., 2009). However, differing allocation methods can represent a political challenge, because they inflict distributional concerns with stakeholders (Santikarn et al., 2018, Burtraw et al., 2013, Flachsland et al., 2009a).

- *Measurement, reporting and verification (MRV), registries, enforcement and market oversight*: MRV is the central activity for ensuring that an ETS is effective, covered entities comply and that the responsible governmental authority controls the carbon market. MRV standards are important for the integrity and stability of emission trading schemes (Edenhofer et al., 2007, Flachsland et al., 2009a). Such provisions include an *emissions registry*, which, if consolidated among linked systems, can improve the transparency of transactions and reduce errors such as double counting (Edenhofer et al., 2007). MRV also include compliance and enforcement mechanisms, e.g. penalty fees for those companies that do not deliver allowances at the end of a commitment period or report false data; these compliance tools can increase the confidence of market participants (Tuerk et al., 2009). The *market oversight* can be seen as the controlling entity that ensures the safety and integrity of the ETS through ETS infrastructure measures such as identity checks, robust software interfaces and measures to prevent cyberattacks (Kachi and Frerk, 2013). The ETS operates based on *compliance periods*, which determine the timeframe for participating entities to submit an allowance per emitted ton of GHG. Generally, these ETS compliance and controlling provisions do not have to be fully aligned for linking, but they have to be at least be comparable across systems and trustworthy (Burtraw et al., 2013), since they are crucial for the integrity and guarantee the stringency of the ETS (Tuerk et al., 2009, Flachsland et al., 2008). They should be recognized by the linking partners as robust, credible and transparent (Flachsland et al., 2008, Haites and Wang, 2009, Tuerk et al., 2009).

- *Cost containment measures*, such as a *price floor* or a *price ceiling*, are designed to hold the carbon price within a politically determined price window. They represent a challenge to linking, because they are difficult to harmonize, as they reflect the political objectives and priorities an authority has set and negotiated (Burtraw et al., 2013, Kachi et al., 2015, Santikarn et al., 2018, Zetterberg, 2012, Hawkins and Jegou, 2014, Haites, 2014, Ranson and Stavins, 2016, Edenhofer et al., 2007, Ellis and Tirpak, 2006). These measures directly affect the carbon prices and thus interfere with domestic price policy objectives (Edenhofer et al., 2007, Ranson and Stavins, 2016, Hawkins and Jegou, 2014, Zetterberg, 2012, Burtraw et al., 2013, Ellis and Tirpak, 2006). For example, if one system implements a carbon price floor, but the other does not, linking may very likely render the price floor obsolete, especially if the system without price containment mechanism features a lower price than that price floor.
- *Flexibility mechanisms* allow companies to *bank* previous allowances or to *borrow* from future trading periods during different phases of compliance. If systems have not aligned their provisions for banking through linking, the combined system will effectively revert to the more generous rules (Tuerk et al., 2009). Various authors find that banking provisions are rather unproblematic for linking, because they likely have little effect on the overall performance of the market (Burtraw et al., 2013, Sterk and Schüle, 2009, Jaffe and Stavins, 2007).
- *Offsets* are external emissions reduction credits accepted by the ETS which have been generated through carbon-offsetting activities outside the ETS, for example through CDM projects. Authors have also discussed that in a linked market different *offset rules* become obsolete, because all (kinds of) permits are available to all participants (Burtraw et al., 2013, Haites and Mullins, 2001, Sterk and Schüle, 2009, Tuerk et al., 2009, Zetterberg, 2012, Hawkins and Jegou, 2014). This situation would threaten political priorities of some ETSs, such as the wish to exclude certificates originating from carbon sinks or other sectors or attempts to limit the amount of offset certificates flowing into the ETS.

- The *scope and coverage* of the ETS describe the sectors of an economy, such as energy, industry, agriculture, waste management, transport or forestry, as well the different GHG gases, that are covered. It is closely related to the *point of regulation*, which defines whether gases are regulated at the production or consumption side of an economy. Most experts have come to the conclusion that the coverage of different sectors or gases does not pose a risk to linkage (Ellis and Tirpak, 2006, Sterk et al., 2006a, Sterk and Schüle, 2009, Haites and Mullins, 2001), except for a situation where a jurisdiction excludes certain sectors for political reasons (Metcalf and Weisbach, 2011).

In principle, linking between systems with similar design and policy objectives would be the easiest approach. Nevertheless, in practice, ETSs differ significantly in their design. This is a very short glimpse into the academic discussion and many more important aspects of linking could be named here. Table 2.1 summarizes the design elements and their relevance for linking, based on a literature assessment of the author.[1]

This section of the book can only hint at the technical complexities involved in linking. At the same time these aspects already reveal that behind many technical challenges stand political obstacles, political priorities, or structural elements of the ETS operating jurisdiction's political system. This further underlines the core assumption of this book that the political conditions determine whether a linking attempt will be successful or not. The following section of this chapter will provide an international relations and international climate policy perspective on linking and presents a set of conditioning factors to linking.

Table 2.1 ETS design elements and their relevance for linking

ETS Design Element	Relevance for Linking
Carbon or allowance price	*relevant*
Emissions cap	*relevant*
Allocation of allowances	*relevant*
MRV and enforcement	*very relevant*
Market oversight	*relevant*
Compliance periods	*less relevant*
Cost containment measures: price ceiling, price floor, reserves	*very relevant*
Flexibility measures: banking and borrowing	*less relevant*
Offset regulations	*very relevant*
Scope and coverage	*less relevant*

Source: Author

Looking at linking with an international relations theory lens

ETS linking can be regarded as a form of cooperation. In the process of linking, two or more actors engage in a mutually beneficial agreement and coordinate their respective ETS policies. In other words, they agree on a common and coordinated policy. In the field of international relations, cooperation is among the oldest and most frequently discussed topics. Cooperation can be found '(...) when actors adopt their behavior to the real or expected interests and preferences of other actors through mutual policy coordination (...)' (Keohane, 1984: 51f). Milner (1992: 468) takes cooperation as goal-directed behavior, where actors mutually receive rewards through adjusting their policies. The author points toward two essential assumptions of this definition: first, that actors make rational choices in order to achieve a certain goal and, second, that the cooperation will provide the actors with some benefits or rewards (Milner, 1992). Thus, this first assumption recognizes that cooperation involves a way of 'behaving' for actors, which can be nation states, institutions, groups or individuals. The second assumption establishes that cooperation happens only if cooperating partners expect gains. Following this idea, also ETS linking requires mutual decisions and collective action.

ETS linking takes one (or several) strategic decisions, where expected benefits are weighed against potential disadvantages and negative impacts (Flachsland et al., 2009a, Haites, 2014). Following rational choice thinking, linking will be more likely to occur if the expected gains from the link surpass its potential costs. An evaluation is carried out by each party to assess whether its domestic political priorities will be supported or undermined by the link. Linking can be seen as a problem of collective action (Interview 8). In the event of linking, the collective outcome (i.e. a link) would be beneficial; however, actors will not cooperate so long as their individual benefits do not outweigh the costs.

The decision-making and negotiation process for linking takes place on multiple levels. It is not simply a *two-level* (Putnam, 1988) but rather a *multilevel* game (Milner, 1992). According to Putnam (1988), negotiations occur at the international level, but also at the domestic level, where the agreement must be ratified. Domestic interests are brought into the international level, where negotiators, for Putnam, governments, intend to form the negotiated deal in a way that satisfies these domestic interests as much as possible. In general, linking implies multilevel governance.

So, what determines whether a cooperation is realized? According to Moravcsik (1997), states' behavior in world politics and cooperation depend on the preferences of each state. These preferences are built on the sum of the state-society relationship, as well as its interdependence structures with other states. *Domestic interests*, but also other domestic *structural aspects* (Risse-Kappen, 1991, Risse-Kappen, 1994) and further (*international*) *systemic factors* (Keohane, 1984) establish the conditions for cooperation. The derivation of states' interests is central to explaining international relations (Moravcsik, 1997). Frieden (1999) refers to national or subnational interests as 'preferences' for specific outcomes, such as wealth. Domestic interests are reflected in the states' positions and brought to the discussion between international delegations, where an agreement is negotiated. Some international relations scholars see domestic political structures as the main factor behind a state's behavior in world politics. Risse-Kappen (1994) offers a definition of such domestic structures: 'Domestic structures encompass the organizational apparatus of political and societal institutions, their routines, the decision-making rules and procedures as incorporated in law and custom, as well as the values and norms prescribing appropriate behavior embedded in the political culture' (Risse-Kappen, 1994: 209). Additionally, international relations authors argue that both the international regime, including its structures and setting of norms, rules, and procedures, as well as the priorities exercised by the other jurisdictions, are part of the explanation for why jurisdictions cooperate (Keohane, 1984, Moravcsik, 1997, Milner, 1992, Frieden, 1999).

The following section of this book extracts factors that have been identified as impediments to or drivers of cooperation by the academic debate on international relations and international environmental and climate policy. Clustered around *domestic interests*, *domestic structures*, and *international developments*, it connects these considerations to the linking debate in order to systematize the influencing factors for linking. These factors build up to a framework for analysis of the political conditions of linking (see Figure 2.1).

Economic interests: The fulfillment of economic interests is the parameter that can be found in most approaches to international cooperation, independent of the authors' school of thought or thematic content. Actors cooperate in order to realize absolute or/and relative gains, usually expressed in economic net benefits: cost reductions, or other financial rewards, such as the improvement of trading conditions (Milner, 1992). In international climate policy, cooperation takes place in order to achieve competitive advantages (Keohane and Victor, 2016),

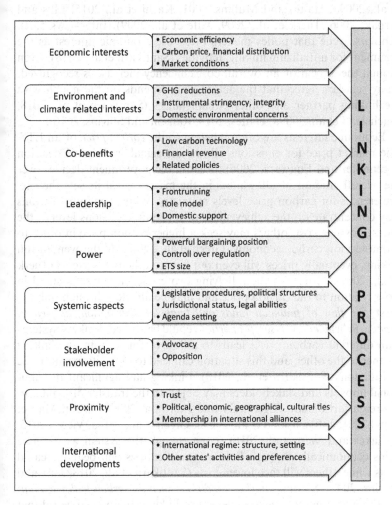

Figure 2.1 The political conditioning factors for linking.
Source: Author

to achieve technological advancement (e.g. renewable energies and low carbon technologies), to share the costs for action or to avoid future costs through climate change. However, it is the potential *economic efficiency* gain expected through linking which functions as a driver for jurisdictions to link. This gain validates the basic idea of linking by showing that linking may lower the overall costs of entities involved in the ETS (Santikarn et al., 2018, Burtraw et al., 2013, Flachsland et al., 2009a, Haites and Mullins, 2001, Kachi et al., 2015, Ellis and Tirpak, 2006, Tuerk et al., 2009, Jaffe et al., 2009) (Interviews 5, 7). Scholars argue that policy-makers often use economic interest as the argument for initializing linking activities (Santikarn et al., 2018). Even though the appeal of an overall cost efficiency increase is recognized, many scholars agree that the actual gain depends on the match with the linking partner and other circumstances (Santikarn et al., 2018, Flachsland et al., 2009a, Haites, 2014, Ranson and Stavins, 2016).

Economic interests are often inherent to the *carbon price* of an ETS. The market price for emissions allowances signals a political decision (Egenhofer and Fujiwara, 2006), because the government decides over the overall amount of allowances in the market when its sets the cap. Preferences for carbon price levels may vary. While some jurisdictions may concentrate on the achievement of a certain emissions level at the lowest possible cost, others may seek a higher carbon price in order to promote a low carbon economy (Green et al., 2014b). In the event of full linking, eventually prices will even out between the two systems (Tuerk et al., 2009). Moreover, the following two economic interests stand in close relation to the carbon price and may be affected through linking. First, the *flow of financial funds and financial distribution, as well as changes in 'winner and loser' proportions* could be altered. Linking systems with different carbon prices leads to a flow of funds from one linking partner to the other, and this situation can lead to opposition within the jurisdictions (Santikarn et al., 2018). Linking has significant distributional impacts and stakeholders may object to the transfer of capital to a foreign jurisdiction (Egenhofer and Fujiwara, 2006, Carbon Market Watch, 2015a, Flachsland et al., 2009a, Green et al., 2014a, Victor, 2015, Jevnaker and Wettestad, 2016). Even though the system as a whole gains economically through linking, some entities will bear increased costs while others will face lower costs (Santikarn et al., 2018, Jaffe and Stavins, 2007). Second, *market conditions such as carbon leakage, competitive issues and trading* are relevant to linking. In an ideal world that moves towards a global carbon price through ETS linkage, competitive concerns would be leveled (Santikarn et al., 2018, Flachsland et al., 2009a). As a result, companies would have less motivation to move production to other world regions, thus alleviating carbon leakage concerns. Industrial groups in particular have an interest in seeing carbon leakage alleviated, which may motivate their support for linkage (Interview 5).[2] However, the expectation of a higher carbon price after linking could also renew the fear for increased carbon leakage, thereby diminishing industrial groups' interest in linking. The improvement of competitive distortions can also be motivator for linking (Tuerk et al., 2009,

Jevnaker and Wettestad, 2016) (Interview 12). A small ETS especially may be interested in alleviating competitive concerns by gaining access to a larger market (Jaffe and Stavins, 2007, Tuerk et al., 2009).

Environmental and climate policy related interests: The generation of a public good—here a healthy atmosphere—can be seen as the ultimate goal of climate-political cooperation. The overall goal of coping with climate change is regarded as a domestic interest, which drives actors to agree on international climate treaties (Keohane and Victor, 2016). An ETS is implemented with specific climate, energy and environmental policy objectives, first and foremost of these being the *reduction of GHG emissions*. Linking of ETSs could in principle improve climate policy, for instance through cost reductions (linking could theoretically achieve the same mitigation at a lower cost), thereby incentivizing policy-makers to commit to stronger climate actions (Borghesi and Montini, 2014, Santikarn et al., 2018, Green et al., 2014a, Mehling et al., 2018). Aside from such theoretical environmental benefits, several authors argue that one of the most serious challenges for linking partners is the coordination on domestic climate policy objectives and ambition levels (Carbon Market Watch, 2015a, Green et al., 2014a, Santikarn et al., 2018) (Interviews 4, 9). To drive linking, climate policy objectives and ambition levels should be similar among the partner jurisdictions (Ranson and Stavins, 2016) (Interviews 16, 8, 7).

Inherent to environmental interests such as GHG reductions or climate policy ambition is a politically decided level of *instrumental stringency and environmental integrity*. A jurisdiction can implement an ETS with stringent or less stringent rules. These can be seen in several design aspects, such as the cap, market control, compliance mechanisms, MRV and offsets utilization (Tuerk et al., 2009, Kachi et al., 2015).[3] Ultimately, linking systems with differing levels of stringency will be politically challenging, because such discrepancies are the result of different political contexts and levels of political support for an ETS (Interview 2). The stringency shows how credible and reliable a system is to others and indicates its overall environmental integrity. The ability of its operation to provide some sort of environmental service, such as emissions reductions, refers to its degree of environmental effectiveness. A jurisdiction with a strong climate-political ambition is likely to establish a stringent system that seeks to guarantee environmental integrity in order to obtain major emissions reductions. Problems of stringency or integrity often originate from the oversupply of allowances in an ETS; for example, if ETS participants receive more free allowances than their effective GHG emissions are; in the situation where different or incoherent GHG calculations stand behind one unit of CO_2; or if market actors defraud market regulations as a result of weak compliance controls or missing market oversight (Kachi et al., 2015). Systemic problems, such as the lack of stringency or integrity, are also likely to persist after linking (Carbon Market Watch, 2015a, Green et al., 2014a, Ranson and Stavins, 2016, Flachsland et al., 2009a).

Contrary to the initially displayed argument that linking may strengthen climate policy, there can be further negative environmental and climate consequences associated with linking (Gulbrandsen et al., 2019, Green, 2017). Certain circumstances, such as the lack of stringency in one of the ETSs, could lead to a situation in which the linked system does not achieve the same level of emissions reductions as both individual jurisdictions would have without linking. Such a scenario potentially threatens domestic reduction targets (Jaffe and Stavins, 2007, Comendant and Taschini, 2014, Carbon Market Watch, 2015a, Green et al., 2014a, Green, 2017, Ranson and Stavins, 2016, Flachsland et al., 2009a). Poor environmental integrity and stringency can also raise issues of equity. Interest groups who see their ETS as more stringent or ambitious than the partner's ETS may perceive the linked system as unfair (Flachsland et al., 2009a). Companies in the more 'lax' ETS will benefit from the reduction efforts of the more stringent ETS, generating a free rider problem (Green et al., 2014a, Flachsland et al., 2009a). The decision to link is also tied to gaming (Green et al., 2014a). Policy-makers may have an incentive to relax the emissions cap over time, thereby weakening emission targets as compared to a scenario without linkage (Haites, 2014, Hawkins and Jegou, 2014, Comendant and Taschini, 2014).

Only very few academic contributions, e.g. Gulbrandsen et al. (2019), examine whether jurisdictions have an interest for *emissions reductions to occur in specific domestic settings*. However, the interviews realized for this work revealed that the focus on domestic emission reductions is a relevant factor in the linking discussion (Interviews 2, 14, 19). Jurisdictions tend to prefer that at least some of their emissions reductions to be realized within their territory. This can be expressed either directly, in a climate target that has to be achieved domestically, or indirectly, in the ETS policy, for example if the ETS does not allow foreign offsets. Linking affects this interest, because under a linked market, emissions will not necessarily be reduced locally. Therefore, a local reduction objective could be undermined by linking.

Co-benefits: Many scholars have noted that the endeavor of cooperation can bring additional advantages, so called 'co-benefits'. Conceptualized as 'side-payments', these can be understood as additional benefits that the negotiator promises or expects in combination with some cooperation deal, and which help to convince constituents to support said cooperation deal (Putnam, 1988). This is also referred to as issue linkage: Cooperation is achieved more easily when several thematic areas are combined (Scott et al., 1995, Haas, 1980). Overall, co-benefits refer to a win-win strategy that pursues more than one objective with only one measure (Mayrhofer and Gupta, 2016); for example, pursuing both climate change mitigation and air quality improvement through an air pollution reduction measure (Nemet et al., 2010). ETSs are also implemented with certain expectations of side-payments or co-benefits. First, ETSs could spur innovation and development in *low carbon technology*, paving the way towards a low carbon economy (Eden et al., 2016). Second, in similarity to a carbon tax, many ETSs are able to *generate financial revenue* by auctioning a share of their emissions allowances. This can become a significant additional source of financial revenue for the government which, depending on the jurisdiction's ETS regulation, can be designated to a variety of uses; such as the funding of further climate protection activities, tax reductions, or increased public assistance for disadvantaged groups (ICAP, 2019b). Third, ETSs can impact a wide variety of *related policies and programs*, such as eco taxes, energy efficiency measures, green jobs, public health protection and energy costs (Santikarn et al., 2018). ETSs are often strongly related to other environmental goals, such as the reduction of air pollution. In some ETSs these local goals even have a stronger character than the global objective (Gulbrandsen et al., 2019). By altering the regional focus of abatement activities, linking can affect the generation and distribution of co-benefits that come with ETS implementation (Tuerk et al., 2009). Linking could both increase or decrease the expectation of certain co-benefits.

Leadership: Among the most discussed factors to motivate environmental cooperation is leadership. Leadership refers to a situation in which one actor advances an ambitious activity or position and brings others to join this activity (Keohane and Victor, 2016). One country, through its dominant responsibility or capacity to address the problem, can lead by providing the largest share of action on climate change (Keohane and Victor, 2016). This kind of situation could be tied to structural leadership, where the leader influences other countries through power, stemming (mainly) from material and economic capacity (Grubb and Gupta, 2000, Gupta and Ringius, 2001). Leadership is also motivated by the expectation of reputational benefits or losses and by 'being perceived as leader by others' (Keohane and Victor, 2016). 'Directional leadership' refers to a situation, where the leader shows through domestic implementation that an objective is achievable. Such '*frontrunning*' raises a normative 'standard' that others feel compelled to abide by (Grubb and Gupta, 2000, Gupta and Ringius, 2001), in order to avoid stigmatization as laggards (Keohane and Victor, 2016). In the process of policy diffusion, one country or region implements an environmental policy, or sets a standard, by unilaterally adopting a policy. The expectation of obtaining leadership role through linking and the demonstration of a successfully linked market could act as a motivator for a linking process (Carbon Market Watch, 2015a, Flachsland et al., 2009a). Linking would demonstrate that this specific ETS is a *role-model* and good or feasible, because other regions choose to connect with it. Linking justifies the interest in promoting and transferring the domestic ETS model to other regions (Interviews 14, 15) and, in principle, it could encourage other regions to install ETSs and trigger further links. Linking could help spread an ETS model to a larger region and further jurisdictions, forming blocks or 'clubs' of jurisdictions as part of a political strategy (Victor, 2015). Linking has also a *symbolic value*: domestically, through creating *more support and awareness of ETSs* and a momentum for action on climate policy; and internationally, as linking can *signal commitment* to international climate policy (Santikarn et al., 2018).

Power constellations and ETS size: Leadership is also strongly related to power, for example, when one ETS operating jurisdiction functions as standard setter for others, thereby gaining followers. In this context, power has been defined as the position or state of capability to pressure international counterparts in negotiations or actors in the domestic policy process (Putnam, 1988). What can be extracted from international relations theory is the notion that power serves as bargaining leverage for cooperation. Constellations of bargaining power are relevant, because they weigh in domestic constraints and opposition, as well as responsibility for a problem (Putnam, 1988, Thomson, 2010). Countries' respective shares of global emissions, for example, were an important intervening variable in the international negotiations on a global climate agreement in Kyoto (Thomson, 2010). In such negotiations, countries with larger shares of emissions, like the United States and China, have more bargaining power; this is due to the fact that their actions would have larger impacts, and their participation is indispensable for solving the global climate change problem (Terhalle and Depledge, 2013).

Holding a *powerful bargaining position*, and thus being able to influence policy and politics, plays a strong role in linking. It is fair to assume that in the case of an ETS, as for all policies, responsible authorities are interested in maintaining control and decision-making power over the system. Linking can therefore threaten these interests because, like other forms of cooperation, it implies sharing of responsibilities. Thus, linking always implies a certain *loss of control over regulation* (Edenhofer et al., 2007), targets and the operation of the system (Ranson and Stavins, 2016). The linked market requires the sharing of control and increased transnational coordination (Flachsland et al., 2009a, Carbon Market Watch, 2015a, Victor, 2015, Comendant and Taschini, 2014, Green et al., 2014a, Ranson and Stavins, 2016). In a linked ETS many decisions and day-to-day routines must be coordinated through constant exchange between responsible agencies on the technical level. Both jurisdictions will be somehow influenced by external ETS developments in the other jurisdiction; these may include economic crises or sudden price shocks, but also directly related policy decisions, such as those related to climate or energy policy (Green et al., 2014a, Wettestad and Jevnaker, 2014, Ranson and Stavins, 2016). Moreover, problems experienced in one ETS may be imported by the linking partner (Wettestad and Gulbrandsen, 2013, Wettestad and Jevnaker, 2014, Tuerk et al., 2009, Flachsland et al., 2009a) (Interviews 8, 7, 3). The capacity to dominate is also associated with structural elements of power, such as the economic strength of the ETS governing jurisdiction, its knowledge capacity and ultimately the *size of the ETS*, that is, the amount of covered emissions. Some authors regard the size of the ETS as the most important influencing variable for linking (Doda and Taschini, 2017). The more GHGs the ETS covers, the stronger the jurisdiction's negotiation position. This is because, in theory, a larger ETS has more effect on global emissions reductions, and will also more strongly dominate the linked market. From an economic perspective a larger ETS can be more attractive, because it offers more trading opportunities and better market functioning for factors such as price stability. It is also suggested that ETSs that are significantly larger than their potential partners are less concerned about potential linking pitfalls, because they only expect a small impact on their system (Santikarn et al., 2018).

Systemic influences and agenda setting: Some authors have noted that specific characteristics of the domestic political system can have an influence on states' behavior in cooperation, such as the legislative procedures, the decision-making rules and law (Risse-Kappen, 1991). For example, the legal procedures for the ratification and conclusion of an agreement play a role, because factors such as the type of parliamentary or congressional majority that is needed affect the availability of options that guarantee domestic adoption (Putnam, 1988). Such structural aspects have been neglected so far in the ETS linking literature. *Legislative procedures* are important in ETS decision making. The process of linking takes place in the context of at least two domestic political systems, whose characteristics and developments will influence political decisions and behaviors. Linking can be influenced by *changing political structures and developments*, such as elections and *government turnover, changes of legislative majorities*, as well as the *overall formal and informal legal abilities of the jurisdiction*. For example, the prospect of a government turnover may bring linking negotiations to a stop if policy-makers expect a change in ETS policy (Interviews 18, 2, 9, 16, 11).

A further aspect that has been reviewed in linking literature is the different *legal statuses of the ETS operating jurisdictions*. The fact that some ETSs are regulated by subnational governments limits their ability to enter into international treaties (Haites and Mehling, 2009, Lenz et al., 2014, Mehling, 2016). Legal compatibility is important for linking and may affect policy-makers linking decisions (Ranson and Stavins, 2013).

But also more informal procedures can be relevant. For example, the set of actions and procedures prior to the implementation and ratification process, such as *agenda setting*. Some discussion has focused on why certain issues rise to and others fall from the political agenda (Anthony, 1972) and who brings which issues onto the agenda, and in what way, e.g. the role that the public and the media play (Weaver et al., 1981). A main observation is that issues often rise or fall on the political agendas independent of the perceived urgency of the topic (Anthony, 1972, Baumgartner and Jones, 2010). Linking can be motivated through a *prioritization on the political agenda or the lack of competing topics*. Generally, linking is benefitted when climate protection is high on the political agenda (Interview 12). Yet, linking is only one among many climate and other topics governments and their administrative bodies act on and it must compete for prioritization on the political agenda.

Stakeholder involvement—advocacy and opposition: Neoliberal theory suggests that the interdependence of governments with political parties, as well as with other actors such as banks, companies, civil actors, NGOs and international organizations, plays a significant role in international cooperation (Frieden, 1999, Moravcsik, 1997). Neoliberal scholars also break with the realist thinking of states as unitary actors, seeing them rather as the sum of the governmental and further actors (Moravcsik, 1997). Political decisions emerge from the bargaining between bureaucracies, organizations and other actors, like interest groups (Mintz and DeRouen Jr, 2010), and shape and mold the national interest in whichever ways also maximize their own organizational benefit (Checkel, 2008).

The implementation and the design of an ETS are the result of a negotiation process between domestic stakeholders (see for example the implementation of RGGI (Rabe, 2008)). Often, the introduction of an ETS causes opposition from domestic stakeholder groups opposed to expected impacts, such as perceived economic, environmental, and social risks. Domestic opposition to cap-and-trade policies may be a severe obstacle to linking, as linking depends on the existence of domestic political support (Ranson and Stavins, 2016, Jevnaker and Wettestad, 2016).

Stakeholders may actively *oppose or advocate* linking, depending on the heterogeneous interests of stakeholders, their capability in terms of knowledge and resources, and the influencing channels existing in the jurisdictional system. Accordingly, they are liable to take action, such as through lobbying, influencing public debates, and even launching lawsuits against a prospective linkage. The stakeholders affected through linking within their respective sector are: *corporate sector*: companies directly regulated by the ETS, indirectly affected companies, service providers; *civil society*: NGOs, academia, and other civil society groups are often more indirectly affected through an ETS and linking through for example, rising energy prices; and *governmental and other political actors*: regional and local governments, authorities, agencies, and political parties, e.g. the ministries for the environment, economy, and finances, who while not directly affected still have an interest in the operation of the ETS policy and potentially a linked market. Stakeholders can influence the linking process at all stages, from before the official start of negotiations until the very last domestic adoption of a linking agreement.

Proximity—culture, trust, diplomatic relationship: Several authors have found parameters that are related to the proximity between partners to be of relevance for cooperation. Such factors could be political culture and shared identity, values and norms (Risse-Kappen, 1991), as well as the existence of historical political relationships. Coincidence, i.e. close ties in such aspects, can facilitate cooperation. A less tangible form of proximity is *trust*. Theoretically, trust has been conceptualized as the willingness to take the risk of putting one's own interests under the control of other actors as well as the expectation that other actors respect and are committed to certain actions (Hoffman, 2002). Trust is discussed as both precondition and result of cooperation (Gambetta, 1988). Within the linking literature, trust, as well as proximity, are usually seen as driving factors (Santikarn et al., 2018, Victor, 2015). *Proximity* in this case refers to *political, economic, geographical*, and *cultural* (such as language or value based) *ties* that have existed previously to the initiation of the linking activity (Jevnaker and Wettestad, 2016). Such proximity often exists between jurisdictions with a common history of cooperation on other issues (Ranson and Stavins, 2013), e.g. border and trade agreements. Observing the 'linking landscape', Ranson and Stavins (2013: 6) even see geographic proximity as the 'single most significant predictor of two systems linking'. An economic argument says that proximity reduces transaction costs. Ranson and Stavins (2013) argue that 'distant' linking partners face higher costs, such as administrative costs, travel costs, and information sharing costs. Various international organizations also include the objective of promoting linking and the establishment of carbon pricing worldwide. *International alliances*, such as ICAP, IETA, CPLC, and the World Bank, have strengthened exchange and trust between jurisdictions implementing or planning to implement carbon pricing instruments worldwide.[4] They provide opportunities for networking and exchange, supporting the learning between the ETS operating jurisdictions and fomenting linking activities more or less directly.

International developments—preferences and setting: International relations authors argue that the *international regime*; its structures and setting of norms, rules and procedures; and the priorities exercised by jurisdictions all help explain why jurisdictions cooperate (Milner, 1992, Keohane, 1984, Frieden, 1999, Moravcsik, 1997). Academic discussion has explored the notion that how other states and regions act on climate change mitigation and their preferences also matter to cooperation partners. This idea is underpinned by the above described notions of power, capability and leadership of certain actors, and also the global nature of the problem; in order to fight climate change effectively, it is essential to get a critical mass of GHG emitters on board. The global spreading of ETS and market-based climate policy tools is often considered to be a driver behind linking, as it creates a generally positive environment for market-based policies. Behind this argument stands the belief that the larger the offer of ETSs worldwide, the higher the chance that an ETS operating jurisdiction will find a linking partner. Also, the opposite has been noted; the still low availability of ETSs worldwide diminishes the chances of linking, because fewer partners are available (Interviews 16, 4, 3, 2). The essential point here is that if new ETSs are developed, they bring alternative linking partners into play. Conversely, the withdrawal of ETS programs and planning can restrict linking opportunities. Both the creation and the withdrawal of ETS policies alter the playing field for potential linking partners and can drive or prevent potential linking partners from connecting. In addition, also *international structures*, for example legal frameworks such as the Kyoto Protocol or the Paris Agreement, are part of the conditions for linking.

Notes

1 The table is based on a rough assessment of secondary literature and linking studies, such as referred to in the text. No quantitative criteria or other evaluation tools were applied. Also, it can be assumed that in each individual linking case the relevance of ETS design elements will be different.

2 It is noteworthy that in practice it is very complicated to calculate the exact price effects of linking and their impact on economic interests. Therefore, it is rather tendencies and expectations, such as a price increase expected by policy-makers or other stakeholders, that act as motivators or barriers for a prospective link.

3 One way that stringency can be measured is to calculate the extent to which the number of allowances represent a reduction from historic or business as usual (BAU) emissions, also known as the marginal cost of compliance. Nonetheless, it is very difficult to compare efforts in numbers, and full comparability at the sectorial or installation level for linking partners is very unlikely to be achieved (Goers et al., 2012, Blyth and Bosi, 2004).

4 ICAP 'facilitates cooperation between countries, sub-national jurisdictions and supranational institutions that have established or are actively pursuing carbon markets through mandatory cap and trade systems' (https://icapcarbonaction.com/en/partnership/about); CPLC 'brings together leaders from government, business, civil society and academia to support carbon pricing, share experiences and enhance the global, regional, national and sub-national understanding of carbon pricing implementation' (www.carbonpricingleadership.org/who-we-are); IETA is a business organization launched to 'establish a functional international framework for trading in greenhouse gas emission reductions' (www.ieta.org/About-IETA).

References

ANGER, N. 2008. Emissions trading beyond Europe: Linking schemes in a post-Kyoto world. *Energy Economics*, 30, 2028–2049.

ANTHONY, D. 1972. Up and down with ecology–The 'issue-attention cycle'. *National Affairs*, 28, 38–50.

BARON, R. & BYGRAVE, S. 2002. *Towards international emissions trading: Design implications for linkages* [Online]. Available: www.oecd.org/environment/cc/2766158.pdf [Accessed 17.04.2019].

BAUMGARTNER, F. R. & JONES, B. D. 2010. *Agendas and instability in American politics*, Chicago: University of Chicago Press.

BLYTH, W. & BOSI, M. 2004. *Linking non-EU domestic emissions trading schemes with the EU emissions trading scheme* [Online]. IEA, OECD Information Paper. Available: www.iea.org/textbase/papers/2004/non_eu.pdf [Accessed 13.12.2014].

BORGHESI, S. & MONTINI, M. 2014. *Linking emission trading schemes around the world: Critical analysis and perspectives* [Online]. Available: http://fessud.eu/wp-content/uploads/2015/03/Linking-Emission-Trading-Schemes-around-the-world-critical-analysis-and-perspectives-working-paper-86.pdf [Accessed 17.04.2019].

BURTRAW, D., PALMER, K. L., MUNNINGS, C., WEBER, P. & WOERMAN, M. 2013. *Linking by degrees: Incremental alignment of cap-and-trade markets* [Online]. Resources for the Future 13-04. Available: www.rff.org/publications/working-papers/linking-by-degrees-incremental-alignment-of-cap-and-trade-markets/ [Accessed 12.03.2017].

CARBON MARKET WATCH. 2015a. *Towards a global carbon market: Prospects for linking the EU ETS to other carbon markets* [Online]. Available: https://carbonmarketwatch.org/wp-content/uploads/2015/05/NC-Towards-a-global-carbon-market-report_web.pdf [Accessed 02.04.2020].

CHECKEL, J. T. 2008. Constructivism and foreign policy. *In:* SMITH, S., HADFIELD, A. & DUNNE, T (ed.) *Foreign policy: Theories, actors, cases.* Oxford: Oxford University Press.

COMENDANT, C. & TASCHINI, L. 2014. *Submission to the inquiry by the House of Commons Select Committee on Energy and Climate Change on 'Linking Emissions Trading Systems'* [Online]. Centre for Climate Change Economics and Policy Grantham Research Institute on Climate Change and the Environment. Available: www.cccep.ac.uk/Publications/Policy/docs/Comendant-and-Taschini-policy-paper-April-2014.pdf [Accessed 20.07.2016].

DODA, B. & TASCHINI, L. 2017. Carbon dating: When is it beneficial to link ETSs? *Journal of the Association of Environmental and Resource Economists,* 4, 701–730.

EDEN, A., UNGER, C., ACWORTH, W., WILKENING, K. & HAUG, C. 2016. *Benefits of emissions trading* [Online]. Berlin: ICAP Policy Paper. Available: https://icapcarbonaction.com/en/?option=com_attach&task=download&id=575 [Accessed 03.04.2020].

EDENHOFER, O., FLACHSLAND, C. & MARSCHINSKI, R. 2007. *Towards a global CO_2 market* [Online]. Available: https://pdfs.semanticscholar.org/29fd/f66e2341bc63d2d57b9b431a42ddf2a328b5.pdf [Accessed 03.02.2020].

EGENHOFER, C. & FUJIWARA, N. 2006. *The contribution of linking emissions markets to a global climate change agreement: Feasibility and political acceptability* [Online]. Brussels: Final report prepared for Environmental Studies Group, Economic Social Research Institute, Cabinet Office, Government of Japan, Centre for European Policy Studies. Available: www.esri.go.jp/jp/prj/int_prj/prj-2004_2005/kankyou/kankyou17/02-1-R.pdf [Accessed 02.04.2019].

ELLIS, J. & TIRPAK, D. 2006. *Linking GHG emission trading systems and markets* [Online]. Organisation for Economic Co-operation Development, International Energy Agency. Available: www.oecd.org/environment/cc/ 37672298.pdf [Accessed 03.03.2019].

FLACHSLAND, C., MARSCHINSKI, R. & EDENHOFER, O. 2009a. To link or not to link: Benefits and disadvantages of linking cap-and-trade systems. *Climate Policy*, 9, 358–372.

FLACHSLAND, C., MARSCHINSKI, R. & EDENHOFER, O. 2009b. Global trading versus linking: Architectures for international emissions trading. *Energy Policy*, 37, 1637–1647.

FLACHSLAND, C., O. EDENHOFER, M. JAKOB & STECKEL, J. 2008. *Developing the international carbon market. Linking options for the EU ETS* [Online]. Report to the Policy Planning Staff at the German Federal Foreign Office. Available: www.mcc-berlin.net/fileadmin/data/pdf/PIK_Carbon_ Market_Linking_2008.pdf [Accessed 03.02.2019].

FRIEDEN, J. A. 1999. *Actors and preferences in the international relations*, Princeton: Princeton University Press.

GAMBETTA, D. 1988. Can we trust trust. *In:* GAMBETTA, D. (ed.) *Trust: Making breaking cooperative relations.* New York: Basil Blackwell.

GOERS, S. R., PFLÜGLMAYER, B. & LUGER, M. J. 2012. Design issues for linking emissions trading schemes—a qualitative analysis for schemes from Europe, Asia and North America. *Journal of Environmental Science and Engineering*, 1, 1322.

GÖRLACH, B., MEHLING, M. & ROBERTS, E. 2015. *Designing institutions, structures and mechanisms to facilitate the linking of emissions trading schemes* [Online]. Berlin: German Emissions Trading Authority (DEHSt). Available: www.dehst.de/SharedDocs/downloads/EN/perspectives/Linking_ report.pdf?__blob=publicationFile&v=3 [Accessed 02.03.2020].

GREEN, J. F. 2017. Don't link carbon markets. *Nature News*, 543, 484.

GREEN, J. F., STERNER, T. & WAGNER, G. 2014a. A balance of bottom-up and top-down in linking climate policies. *Nature Climate Change*, 4, 1064–1067.

GREEN, J. F., STERNER, T. & WAGNER, G. 2014b. *The politics of market linkage: Linking domestic climate policies with international political economy* [Online]. Fondazione Eni Enrico Mattei Nota di Lavoro 64.2014. Available: www.econstor.eu/bitstream/10419/102003/1/NDL2014-064.pdf [Accessed 03.04.2018].

GRUBB, M. & GUPTA, J. (eds.) 2000. *Climate change, leadership and the EU,* Dordrecht: Kluwer Academic Publishers.

GULBRANDSEN, L. H., WETTESTAD, J., VICTOR, D. G. & UNDERDAL, A. 2019. The political roots of divergence in carbon market design: Implications for linking. *Climate Policy*, 19, 427–438.

GUPTA, J. & RINGIUS, L. 2001. The EU's climate leadership: Reconciling ambition and reality. *International Environmental Agreements*, 1, 281–299.

HAAS, E. B. 1980. Why collaborate? Issue-linkage and international regimes. *World Politics*, 32, 357–405.

HAITES, E. 2014. *Lessons learned from linking emissions trading systems: General principles and applications* [Online]. Washington DC: Partnership for Market Readiness (PMR). Available: www.thepmr.org/system/files/documents/PMR%20Technical%20Note%207.pdf [Accessed 09.09.2019].

HAITES, E. & MEHLING, M. 2009. Linking existing and proposed GHG emissions trading schemes in North America. *Climate Policy*, 9, 373–388.

HAITES, E. & MULLINS, F. 2001. *Linking domestic and industry greenhouse gas emission trading systems* [Online]. Margaree Consultants. Available: http://citeseerx.ist.psu.edu/viewdoc/download?doi=10.1.1.512.9323&rep=repl&type=pdf [Accessed 12.12.2018].

HAITES, E. & WANG, X. 2009. Ensuring the environmental effectiveness of linked emissions trading schemes over time. *Mitigation and Adaptation Strategies for Global Change*, 14, 465–476.

HAWKINS, S. & JEGOU, I. 2014. *Linking emissions trading schemes: Considerations and recommendations for a joint EU-Korean carbon market* [Online]. ICTSD Global Platform on Climate Change, Trade and Sustainable Energy Available: www.ictsd.org/sites/default/files/research/linking-emissions-trading-schemes-considerations-and-recommendations-for-a-joint-eu-korean-carbon-market.pdf [Accessed 09.09.2019].

HOFFMAN, A. M. 2002. A conceptualization of trust in international relations. *European Journal of International Relations*, 8, 375–401.

ICAP. 2015b. *What is emissions trading?* [Online]. ETS BRIEF #1 October 2015 Available: https://icapcarbonaction.com/en/?option=com_attach&task=download&id=377 [Accessed 20.08.2019].

ICAP. 2019b. *From carbon market to climate finance: Emissions trading revenue* [Online]. ETS BRIEF #5. Available: https://icapcarbonaction.com/en/?option=com_attach&task=download&id=674 [Accessed 16.03.2020].

JAFFE, J., RANSON, M. & STAVINS, R. N. 2009. Linking tradable permit systems: A key element of emerging international climate policy architecture. *Ecology LQ*, 36, 789.

JAFFE, J. & STAVINS, R. 2007. *Linking tradable permit systems for greenhouse gas emissions: Opportunities, implications, and challenges* [Online]. IETA report. Available: www.belfercenter.org/sites/default/files/publication/IETA_Linking_Report.pdf [Accessed 02.05.2020].

JEVNAKER, T. & WETTESTAD, J. 2016. Linked carbon markets: Silver bullet, or castle in the air? *Climate Law*, 6, 142–151.

KACHI, A. & FRERK, M. 2013. *Market oversight primer* [Online]. International Carbon Action Partnership Policy Paper. Available: https://icapcarbonaction.com/en/market-oversight [Accessed 15.7.2016].

KACHI, A., UNGER, C., BÖHM, N., STELMAKH, K., HAUG, C. & FRERK, M. 2015. *Linking emissions trading systems: A summary of current research* [Online]. International Carbon Action Partnership Policy Paper. Available: https://icapcarbonaction.com/en/?option=com_attach&task=download&id=575 [Accessed 02.05.2020].

KEOHANE, R. O. 1984. *After hegemony: Cooperation and discord in the world political economy*, Princeton, Princeton University Press.

KEOHANE, R. O. & VICTOR, D. G. 2016. Cooperation and discord in global climate policy. *Nature Climate Change*, 6, 570.

LENZ, C., VOLMERT, B., HENTSCHEL, A. & ROßNAGEL, A. 2014. *Die Verknüpfung von Emissionshandelssystemen–sozial gerecht und ökologisch effektiv*, Kassel: Kassel University Press GmbH.

MAYRHOFER, J. P. & GUPTA, J. 2016. The science and politics of co-benefits in climate policy. *Environmental Science Policy*, 57, 22–30.

MEHLING, M. & GÖRLACH, B. 2016. Multilateral linking of emissions trading systems. CEEPR WP 2016-009.

MEHLING, M. & HAITES, E. 2009. Mechanisms for linking emissions trading schemes. *Climate Policy*, 9, 169–184.

MEHLING, M. A. 2016. Legal frameworks for linking national emissions trading systems. *The Oxford Handbook of International Climate Change Law*. Oxford University Press Oxford.

MEHLING, M. A., METCALF, G. E. & STAVINS, R. N. 2018. Linking climate policies to advance global mitigation. *Science*, 359, 997–998.

METCALF, G. E. & WEISBACH, D. 2011. Linking policies when tastes differ: Global climate policy in a heterogeneous world. *Review of Environmental Economics and Policy*, 6, 110–129.

MILNER, H. 1992. International theories of cooperation among nations. *World Politics*, Vol.44, 466–496.

MINTZ, A. & DEROUEN JR, K. 2010. *Understanding foreign policy decision making*, New York, Cambridge University Press.

MORAVCSIK, A. 1997. Taking preferences seriously: A liberal theory of international politics. *International organization*, 51, 513–553.

NEMET, G. F., HOLLOWAY, T. & MEIER, P. 2010. Implications of incorporating air-quality co-benefits into climate change policymaking. *Environmental Research Letters*, 5.

PUTNAM, R. D. 1988. Diplomacy and domestic politics: The logic of two-level games. *International Organization*, 42, 427–460.

RABE, B. 2008. *The complexities of carbon cap-and-trade policies: Early lessons from the states* [Online]. Governance Studies at Brookings Available: www.brookings.edu/wp-content/uploads/2016/06/1009_captrade_rabe.pdf [Accessed 13.03.2020].

RANSON, M. & STAVINS, R. N. 2013. *Linkage of greenhouse gas emissions trading systems: Learning from experience* [Online]. Washington: Resources for the Future RFF DP13-42. Available: www.rff.org/publications/working-papers/linkage-of-greenhouse-gas-emissions-trading-systems-learning-from-experience/ [Accessed 03.09.2017].

RANSON, M. & STAVINS, R. N. 2016. Linkage of greenhouse gas emissions trading systems: Learning from experience. *Climate Policy*, 16, 284–300.

RISSE-KAPPEN, T. 1991. Public opinion, domestic structure, and foreign policy in liberal democracies. *World Politics*, 43, 479–512.

RISSE-KAPPEN, T. 1994. Ideas do not float freely: Transnational coalitions, domestic structures, and the end of the cold war. *International Organization*, 48, 185–214.

SANTIKARN, M., LI, L., THEUER, S. L. H. & HAUG, C. 2018. *A guide to linking emissions trading systems* [Online]. Berlin: Report published by ICAP. Available: www.icapcarbonaction.org/publications [Accessed 03.09.2019].

SCOTT, G. L., REYNOLDS, G. M. & LOTT, A. D. 1995. Success and failure components of global environmental cooperation: The making of international environmental law. *ILSA Journal of International & Comparative Law*, 2, 23.

STERK, W., BRAUN, M., HAUG, C., KORYTAROVA, K. & SCHOLTEN, A. 2006a. *Ready to link up?: Implications of design differences for linking domestic emissions trading schemes* [Online]. Available: https://epub.wupperinst.org/frontdoor/index/index/docId/2495 [Accessed 02.05.2020].

STERK, W., MEHLING, M. & TUERK, A. 2009. *Prospects of linking EU and US emission trading schemes: Comparing the Western Climate Initiative, the Waxman-Markey and the Lieberman-Warner Proposals* [Online]. Climate Strategies. Available: http://climatestrategies.org/wp-content/uploads/2009/04/linking-eu-us.pdf [Accessed].

STERK, W. & SCHÜLE, R. 2009. Advancing the climate regime through linking domestic emission trading systems? *Mitigation and Adaptation Strategies for Global Change*, 14, 409–431.

TERHALLE, M. & DEPLEDGE, J. 2013. Great-power politics, order transition, and climate governance: Insights from international relations theory. *Climate Policy*, 13, 572–588.

THOMSON, A. 2010. Rational design in motion: Uncertainty and flexibility in the global climate regime. *European Journal of International Relations*, 16, 269–296.

TUERK, A., MEHLING, M., FLACHSLAND, C. & STERK, W. 2009. Linking carbon markets: Concepts, case studies and pathways. *Climate Policy*, 9, 341–357.

VICTOR, D. G. 2015. *The case for climate clubs* [Online]. International Centre for Trade and Sustainable Development (ICTSD) Available: www.e15initiative.org/ [Accessed 27.07.2018].

WEAVER, D. H., GRABER, D., MCCOMBS, M. E. & EYAL, C. H. 1981. *Media agenda-setting in a presidential election: Issues, images, and interest*, New York: Praeger.

WETTESTAD, J. & GULBRANDSEN, L. H. 2013. The evolution of carbon trading systems: Waves, design and diffusion. *International Cooperation*, 40.

WETTESTAD, J. & JEVNAKER, T. 2014. The EU's quest for linked carbon markets. *In:* CHERRY, T. L., HOVI, J. & MCEVOY, D. M. (eds.) *Toward a new climate agreement: Conflict, resolution and governance.* London: Routledge.

ZETTERBERG, L. 2012. *Linking the emissions trading systems in EU and California* [Online]. Fores, Swedish Environmental Research Institute. Available: https://fores.se/wp-content/uploads/2013/04/FORES-California_ETS-web.pdf [Accessed 05.05.2019].

3 EU climate policy and the design and development of the EU Emissions Trading System

Setting the scene

After the attempted introduction of a tax on energy and carbon dioxide emissions stagnated into fruitless discussions for close to a decade, for the EU, the Kyoto Protocol represented a window of opportunity to install an ETS. At that time, the decision to implement an ETS on the European level came rather quickly (Wettestad, 2005). The first proposition for an ETS was released in 2001 through the European Commission, and then the European directive on emissions trading was adopted in 2003: 'DIRECTIVE 2003/87/EC OF THE EUROPEAN PARLIAMENT AND OF THE COUNCIL' (EU COM, 2003). The EU ETS is part of the EU's commitment under the Kyoto Protocol in 2005 and, starting after 2020, the Paris Agreement. In 2014, the EU adopted the '2030 climate & energy framework', which highlights the ETS as a cornerstone instrument. The regulation includes a binding, economy-wide target of at least 40% domestic reduction in GHGs by 2030 (compared to 1990). Additionally, the '2050 Roadmap for moving to a competitive low-carbon economy' introduces a number of mid-term GHG reduction targets: 60% by 2040, and 80% by 2050. Subsequently, the EU also decided to pursue climate neutrality (EU COM, n.d. d).

Following the introduction of EU COM's first comprehensive proposal for the design of an ETS in 2001, conflicts soon emerged. Although the adoption of the ETS Directive was quickly realized, consensus building among the European member states proved to be a complicated process (Skjærseth and Wettestad, 2008, Wettestad, 2005). Despite general consensus over the introduction of an ETS, discussions moved to whether the scheme should be voluntary rather than mandatory, as proposed by the EU COM (Christiansen and Wettestad, 2003). Many industrial associations in the EU had overall positive positions on the introduction of an ETS as a climate policy tool, favoring this solution to the introduction of taxes (Christiansen and Wettestad, 2003). Over time, however, industrial positions and lobbying on European climate policy, especially regarding ETS policy, have changed dynamically and become more divergent. For example, critical voices came from the Alliance of Energy Intensive Industries.[1] In contrast, other entities, such as Eurelectic,[2] IBM and PricewaterhouseCoopers, held somewhat more progressive positions, supporting a strong ETS (Fagan-Watson et al., 2015). Interestingly, while environmental groups initially adopted a generally positive stance on the introduction of an ETS, by around the 2000s they had become concerned about loopholes for polluters and the principal idea of buying credits in place of emissions reduction (Wettestad, 2005, Butzengeiger and Michaelowa, 2004). The European Commission (EU COM) pursued a rather ambitious, environmentally progressive approach to ETS policy (Christiansen and Wettestad, 2003) and was a driving actor in the introduction of the ETS (Wettestad and Jevnaker, 2014).

Further, conflicts emerged also over the design of the EU ETS. For example, in which way emissions allowances were going to be distributed to participating entities, and what the level of the national caps defining the overall amount of allowances would be (Orlowski and Gründinger, 2011).

EU ETS implementation has been shaped by serious challenges from the beginning. Many of these were caused by its rather weak policy design, which was a product of the consensus process with stakeholders and between member states. Probably the most critical and long-lasting problem has been the oversupply of emissions allowances in the EU ETS. The overload of allowances on the carbon market dates back to the EU ETS's design phase and was already evident after the first compliance periods. During phase I of the ETS (2005–2007) significant oversupply resulted from overly generous allocation of emissions allowances by European member states to their domestic companies (Schleicher et al., 2015). The national caps were larger than the BAU emissions levels (Ranson and Stavins, 2012). As allowances could not be 'banked' or carried over to the following phases, the carbon price reached zero by the end of the first phase in 2007, rendering the system ineffective. The surplus of allowances on the market increased during phase II (2008–2012) of the EU ETS in 2008 (Schleicher et al., 2015). Official data from the EU COM calculated a surplus of 2.1 billion allowances in 2013, at the beginning of the third ETS phase (EU COM, n.d. c). The extensive use of cheap credits produced under the CDM also proved to be a major weakness (EU COM, n.d. c). The economic recession was an addition cause for trouble (Hermann and Matthes, 2012, Branger et al., 2014). The subsequent debt crisis within the Euro zone led to decreased industrial production and reduced energy use in the EU member states, dropping in 2012 to about 10% lower than the pre-crisis trend levels (Sartor, 2012). As a result of these various problems, the oversupply of allowances remained within the system into phase III (2013–2020) (see Figure 3.1).

Since 2008, carbon prices for EU Allowances (EUAs) in the EU ETS have remained consistently low. Until spring 2018, they always stayed under the reference price of EUR 30 (Elkerbout and Egenhofer, 2017). In the second phase of the EU ETS (2008–2012) prices experienced a very brief peak of 28 EUR in 2008, but then stayed around 13–16 EUR during 2009–2011 and at an average of 7.4 EUR in 2012. For the majority of the third phase of the EU ETS, prices varied from 2.7 to 8 EUR. Only more recently, in fall 2018, did the EU ETS start to experience prices above 20 EUR.

In close relation to the problem of allowances oversupply stands the issue of other policy instruments interfering with the EU ETS. In theory, an overlap with policies such as renewable energy technology measures will decrease the demand, and therefore prices for allowances (Goulder, 2013, Fankhauser et al., 2010). Experts are ambiguous about the degree to which such policies influence the demand for emissions allowances (Ellerman et al., 2016, Martin et al., 2015, Brink et al.,

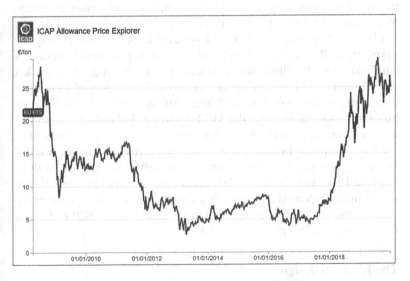

Figure 3.1 Carbon price development in the EU ETS (2008–2019).

Source: ICAP, 2020: Allowance Price Explorer. Retrieved 28.07.2020 from: https://icapcarbonaction.com/en/ets-prices (used with permission granted by the ICAP Secretariat)

2016). For example, experts found that the growth of wind and solar electricity production has had an empirically evident, but only modest impact on EUAs' price drops (Koch et al., 2014).

An additional challenge the EU ETS faced during the first two phases were several cyber-attacks and value-added tax (VAT) frauds. For example, in the case of the company Holcim Ltd., the Romanian registry was hacked and emissions allowances worth EUR 23.5 million were transferred (Interpol, 2013). A culmination of these attacks occurred in January 2011, when around two million EUAs, valued at approximately EU 28.7 million, were stolen, requiring the EU to impose a week-long freeze on spot trading (Reuters, 2011).

Critics also argue that the EU ETS does not provide long-term investment incentives (Elkerbout and Egenhofer, 2017). In other words, it does not spur the necessary investments in technology change and innovation that lead to a decarbonized economy. While this situation is due in part to low carbon prices and weak ETS design choices, the long-term climate targets also play an important role. In 2012, scholars assumed that the emissions targets of ETS (as well as non-ETS sectors) were not in line with the EU long-term GHG reduction strategy (80% reduction in 2050 compared to 1990) (Hermann and Matthes, 2012).

Overall, these challenges have compromised the EU ETS's environmental effectiveness, leading the EU to employ several reform options to enhance the EU ETS over the past decade (Stephan and Sartor, 2013). As a first step the EU COM initiated back-loading. During the years 2014, 2015, and 2016 a total of 900 allowances were temporarily withheld from the allowances auctions (EU COM, n.d. c). A market stability reserve (MSR) that seeks to regulate the quantity of allowances circulating on the market started operation in 2019 (EU COM, n.d.c). Studies found that in 2019, the MSR withdrew 397 million allowances from auction volumes (ICAP, 2020a). In parallel, a structural reform of the EU ETS phase IV for 2021 was negotiated (EU PA, 2017). It is difficult to give a good estimation on how successful the EU ETS will be in phase IV, as this also depends on economic factors, the international situation and updated climate policy targets. In fact, the updating process for the EU's NDC under the Paris Agreement in 2020 includes a new EU-wide GHG reduction target.

The EU ETS design

The primary aim of the EU ETS is "to promote reductions of greenhouse gas emissions in a cost-effective and economically efficient manner" (EU COM, 2003). Low carbon innovation can be seen as a secondary objective in the EU ETS legislation (EU COM, 2003). Within the climate policy mix, the EU ETS is supposed to achieve the largest emissions reductions in combination with other climate policy instruments, such as car emissions standards or specific funding programs. In parallel, many EU member states employ additional national instruments, such as a carbon tax in France or the recently decided carbon tax in Germany (Bundesregierung Deutschland, 2019).

The EU ETS follows a determined reduction pathway by setting a progressively declining cap for absolute emissions reductions (currently, 1.74% and 2.2% from 2021) in order to achieve the EU's climate policy targets. If followed, this pathway would reduce gases from ETS sectors (except aviation) to 43% below 2005 levels by 2030 (EU COM, n.d. a). The EU ETS covers about 45% of the region's GHG, with a current cap of 1,816 MtCO$_2$e in 2020. It includes various sectors: power and heat generation, industrial processes, aviation and several gases (CO$_2$, N$_2$O, and PFCs) (ICAP, 2020a). The EU ETS contains no provisions that maintain the price within a certain corridor, but allowances can be partially (up to 50% of the overall reduction under the EU ETS, in phases I–II) compensated with credits generated through international projects under the CDM (ICAP, 2020a). In the first and second phase of the EU ETS, most allowances (approximately 90%) were allocated for free and through benchmarking, with only some member states deciding to auction a portion of allowances. Currently, in phase III, 57% of the allowances are auctioned over the entire trading period; the power sector is required to purchase 100% of their allowances in auction, while specific benchmarking rules apply to the industrial sector and aviation receives almost all allowances for free (ICAP, 2020a). From 2012 to 2019, the EU generated an income of approximately 54 billion EUR from auctioning allowances (ICAP, 2019b).[3] The EU recommends spending at least 50% of this on further climate mitigation activities (ICAP, 2019b).

The legal basis of the EU ETS is the so called Emissions Trading Directive (Directive 2003/87/EC) (EU COM, 2003), which was reformed in 2009 (Directive 2009/29/EC) and 2014 (EU COM, 2014). Article 25 of the 'ETS Directive' allows for the EU ETS to link with other systems. The regulation originally permitted agreements with third countries listed in Annex B to the Kyoto Protocol that had ratified the Protocol. In 2009, amendments were adopted that expanded the range of potential linking partners to all jurisdictions that fulfills the following conditions: system compatibility such as basic environmental integrity, the mandatory nature of the ETS and the existence of an absolute cap on emissions EU COM (EU COM, n.d. a).

In the event that linking is considered, the process can take place as follows: The EU COM assesses potential linking partners and talks to them informally. It then recommends the opening of negotiations with the specific jurisdiction to the European Council (EU COU) and the European Parliament (EU PA), who are needed to adopt the official negotiation mandate. For this purpose, the EU COM needs to provide an impact assessment of the planned link. Once the mandate is given, the EU COM may start formal negotiations. During the negotiations EU COM must regularly inform the EU COU and the EU PA on their progress and the stage of the linking process. The EU and the partner jurisdiction conclude negotiations with an agreement, such as an international treaty or MoU. In order to be signed and concluded, the EU COU and EU PA have to adopt this decision (Pinkster, 2015).

Since 2010, Directorate General Climate Action (DG CL) is the EU's responsible authority for all climate matters, ranging from bilateral cooperation to the drafting of regulations or impact assessments. DG CL (and formerly the DG ENV) (EU COM, n.d. a) has been a strong supporter of the EU ETS, having an 'entrepreneurial' and, at least initially, driving role (Skjærseth and Wettestad, 2010, Wettestad and Jevnaker, 2014). The degree to which member states are active during linking and ETS decisions also depends on their capacity, interests and organizational structure (EU COM, n.d. b). They are represented in the EU COU. The EU PA has not been portrayed as having had a very active role in linking activities or noticeable positions against or for it (Wettestad and Jevnaker, 2014).

It can be argued that the spread of linking and the very possibility of a large global carbon market have been intrinsically tied to the emergence and development of the EU ETS. The EU COM had stated an OECD-wide carbon market, that could at some point extend globally, as a principal objective (Grubb, 2012, Mehling, 2007, Jevnaker and Wettestad, 2016, Tuerk et al., 2009). Also other communications from the EU COM named developing a worldwide carbon market as the ultimate goal of ETS and describes linking as "highly desirable" (EU COM, 2008a).[4] Initially, the EU concentrated on a geographical expansion of its ETS on the European territory. In 2007, the decision was taken to incorporate the Emissions Trading Directive into the European Economic Area (EEA), meaning Liechtenstein, Iceland and Norway would be integrated into the EU ETS. The EU described these linkages as an 'extension' of the EU ETS (Marr, 2007). However, in the Norwegian example, a simple 'annexation' was met by a significant, politically controversial debate.

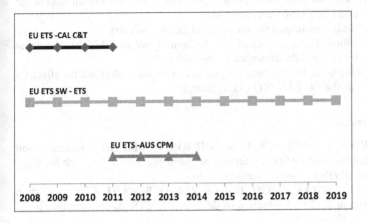

Figure 3.2 Timeline of the EU's linking efforts.
Source: Author

The Norwegian Greenhouse Gas Emission Trading Act of 17 December 2004 established a national ETS (Government of Norway, 2004). Launched in 2005, it was designed very similarly to the EU ETS. However, when the EU and Norway started to negotiate (2005–2006), disagreement appeared on how the link was going to be realized. While Norway wanted to implement the link through a bilateral treaty, the EU COM advocated that Norway, as a member of the EEA, was obliged to adopt the EU Directive (Jevnaker and Wettestad, 2016). The main point of conflict was the required expansion of the ETS to certain sectors, such as the inclusion of offshore installations and the wood processing industry (Mace et al., 2008, Sterk et al., 2006b). Nevertheless, in 2006, Norway agreed to these concessions (Government of Norway, 2006) and joined the EU ETS in 2009 (Jevnaker and Wettestad, 2016).

The following chapters review three specific cases of EU linking efforts. Figure 3.2 gives an overview of the time phases reviewed for this book. The marked rows for California, Switzerland and Australia are the time periods examined in each case.

Notes

1   E.g. the Confederation of European Paper Industries (CEPI), the European Steel Association (EUROFER), Eurometauyx (the trade association for metal producers), the European Chemical Industry Council (CEFIC) and

Business Europe (a lobby group representing enterprises of all sizes in the EU and other European countries).

2   Eurelectric represents the European electricity industry.

3   59 billion USD, converted by the author, as of 03.05.2020 (https://bankenverband.de/service/waehrungsrechner/).

4   Linking is 'highly desirable as long as it does not undermine the effectiveness of the EU ETS' (EU COM, 2008a).

References

BRANGER, F., LECUYER, O. & QUIRION, P. 2014. The European Union Emissions Trading System: Should we throw the flagship out with the bathwater? *WIREs Climate Change*, 6, 9–16.

BRINK, C., VOLLEBERGH, H. R. & VAN DER WERF, E. 2016. Carbon pricing in the EU: Evaluation of different EU ETS reform options. *Energy Policy*, 97, 603–617.

BUNDESRESGIERUNG DEUTSCHLAND. 2019. *CO2-Bepreisung* [Online]. Available: www.bundesregierung.de/breg-de/themen/klimaschutz/co2-bepreisung-1673008 [Accessed 02.05.2020].

BUTZENGEIGER, S. & MICHAELOWA, A. 2004. Greenhouse gas emissions trading in the European Union—Background and implementation of a 'new' climate policy instrument. *Intereconomics*, 39, 116–118.

CHRISTIANSEN, A. C. & WETTESTAD, J. 2003. The EU as a frontrunner on greenhouse gas emissions trading: How did it happen and will the EU succeed? *Climate Policy*, 3, 3–18.

ELKERBOUT, M. & EGENHOFER, C. 2017. *The EU ETS price may continue to be low for the foreseeable future—Should we care?* [Online]. CEPS Policy Insight No 2017/22, June 2017. Available: www.ceps.eu/ceps-publications/eu-ets-price-may-continue-be-low-foreseeable-future-should-we-care/ [Accessed 02.03.2019].

ELLERMAN, A. D., MARCANTONINI, C. & ZAKLAN, A. 2016. The European Union emissions trading system: Ten years and counting. *Review of Environmental Economics and Policy*, 10, 89–107.

EU COM. 2003. Directive 2003/87/EC of the European Parliament and of the Council of 13 October 2003 Establishing a scheme for greenhouse gas emission allowance trading within the community and amending council directive 96/61/EC. European Commission Brussels.

EU COM. 2008a. *Impact assessment. Commission Staff Working Document accompanying document to the Proposal for a Directive of the European Parliament and of the Council amending Directive 2003/87/EC so as to improve and extend the EU greenhouse gas emission allowance trading system* [Online]. Available: https://ec.europa.eu/clima/sites/clima/files/ets/docs/sec_2008_52_en.pdf [Accessed 20.04.2020].

EU COM. 2014. Regulation (EU) No 421/2014 of the European Parliament and of the Council of 16 April 2014 amending Directive 2003/87/EC.

EU COM. n.d. a. *International carbon market* [Online]. Available: https://ec.europa.eu/clima/policies/ets/markets_en [Accessed 20.03.2020].

EU COM. n.d. b. *Effort sharing: Member States' emission targets* [Online]. Available: https://ec.europa.eu/clima/policies/effort_en [Accessed 23.03.2030].

EU COM. n.d. c. *Market stability reserve* [Online]. Available: https://ec.europa.eu/clima/policies/ets/reform_en [Accessed 2.2.2017].

EU COM. n.d. d. *Climate strategies & targets* [Online]. Available: https://ec.europa.eu/clima/policies/strategies_en [Accessed 30.03.2020].

EU PA. 2017. *Post-2020 reform of the EU emissions trading system* [Online]. Available: http://www.europarl.europa.eu/RegData/etudes/BRIE/2017/595926/EPRS_BRI(2017)595926_EN.pdf [Accessed 23.03.2020].

FAGAN-WATSON, B., ELLIOTT, B. & WATSON, T. 2015. *Lobbying by trade associations on EU climate policy* [Online]. University of Westminster. Available: https://westminsterresearch.westminster.ac.uk/item/98xz3/lobbying-by-trade-associations-on-eu-climate-policy [Accessed 09.09.2019].

FANKHAUSER, S., HEPBURN, C. & PARK, J. 2010. Combining multiple climate policy instruments: How not to do it. *Climate Change Economics*, 1, 209–225.

GOULDER, L. H. 2013. Markets for pollution allowances: What are the (new) lessons? *Journal of Economic Perspectives*, 27, 87–102.

GOVERNMENT OF NORWAY. 2004. *Greenhouse gas emission trading act* [Online]. Ministry of Climate and Environment. Available: www.regjeringen.no/en/dokumenter/greenhouse-gas-emission-trading-act/id172242/ [Accessed 23.03.2020].

GOVERNMENT OF NORWAY. 2006. *Norway accept EU emissions trading directive* [Online]. Available: www.regjeringen.no/no/aktuelt/norway-accept-eu-emissions-trading-direc/id419857/ [Accessed 23.03.2020].

GRUBB, M. 2012. Emissions trading: Cap and trade finds new energy. *Nature*, 491, 666.

HERMANN, H. & MATTHES, F. 2012. *Strengthening the European Union emissions trading scheme and raising climate ambition: Facts, measures and implications* [Online]. Report by Öko-Institut. Available: www.wwf.de/fileadmin/fm-wwf/Publikationen-PDF/GP_WWF_2012_-_Strengthening_the_EU_ETS_and_Raising_Climate_Ambition.pdf [Accessed 03.03.2017].

ICAP. 2019b. *From carbon market to climate finance: Emissions trading revenue* [Online]. ETS BRIEF #5. Available: https://icapcarbonaction.com/en/?option=com_attach&task=download&id=674 [Accessed 16.03.2020].

ICAP. 2020a. *Emissions trading worldwide: International carbon action partnership (ICAP) status report 2019* [Online]. Berlin. Available: https://icapcarbonaction.com/en/?option=com_attach&task=download&id=677 [Accessed 02.05.2020].

INTERPOL. 2013. *Guide to carbon trading crime 2013* [Online]. Available: www.interpol.int/Media/Files/Crime-areas/Environmental-crime/Guide-to-Carbon-Trading-Crime-2013 [Accessed 26.3.2015].

JEVNAKER, T. & WETTESTAD, J. 2016. Linked carbon markets: Silver bullet, or castle in the air? *Climate Law*, 6, 142–151.

KOCH, N., FUSS, S., GROSJEAN, G. & EDENHOFER, O. 2014. Causes of the EU ETS price drop: Recession, CDM, renewable policies or a bit of everything?—New evidence. *Energy Policy*, 73, 676–685.

MACE, M. J., MILLAR, I., SCHWARTE, C., ANDERSON, J., BROEKHOFF, D., BRADLEY, R., BOWYER, C. & HEILMAYR, R. 2008. *Analysis of the legal and organisational issues arising in linking the EU Emissions Trading Scheme to other existing and emerging emissions trading schemes* [Online]. Study Commissioned by the European Commission DG-Environment, Climate Change and Air. ENV.C.2/SER/2006/0115r Available: http://static1.squarespace.com/static/56bcccdcb09f954f203561af/t/5720d52df8baf30a23ca975a/1461769520738/SYDDMS-719716-v1-FIELD_EU-ETS_linking_project_2008+(3).PDF [Accessed 23.04.2019].

MARR, S. 2007. *Linking the EU ETS. Opportunities and challenges* [Online]. EU COM, DG Environment. Available: https://ec.europa.eu/clima/sites/clima/files/docs/0068/2a_marr_en.pdf [Accessed 23.03.2020].

MARTIN, R., MUÛLS, M. & WAGNER, U. J. 2015. The impact of the European Union Emissions Trading Scheme on regulated firms: What is the evidence after ten years? *Review of environmental economics policy*, 10, 129–148.

MEHLING, M. 2007. Bridging the transatlantic divide: Legal Aspects of a link between regional carbon markets in Europe and the United States. *Sustainable Development Law & Policy*, 7, 46–51.

ORLOWSKI, M. & GRÜNDINGER, W. 2011. Der Streit um heiße Luft: Der Einfluss von Interessengruppen auf den EU-Emissionshandel und seine Umsetzung in Deutschland und dem Vereinigten Königreich. *Zeitschrift für Public Policy, Recht und Management*, 4, 125–148.

PINKSTER, J. 2015. *EU perspective on linking* [Online]. Ministry of Infrastructure and Environment. Available: http://iki-alliance.mx/wp-content/uploads/2015-01-29_DAY4_Presentation-Pinkster_Linking-in-the-European-ETS.pdf [Accessed 04.04.2020].

RANSON, M. & STAVINS, R. N. 2012. *Post-Durban climate policy architecture based on linkage of cap-and-trade systems* [Online]. National Bureau of Economic Research. Available: www.belfercenter.org/sites/default/files/files/publication/ranson-stavins_dp51.pdf [Accessed 13.03.2020].

REUTERS. 2011. *Timeline: Scandals in the EU carbon market* [Online]. Available: www.reuters.com/article/us-carbon-scandals-idUSTRE70J4M120110120 [Accessed 24.5.2017].

SARTOR, O. 2012. *The EU ETS carbon price: To intervene, or not to intervene?* [Online]. Climate Brief N°12 February 2012. Available: www.i4ce.org/wp-core/wp-content/uploads/2016/03/12-02-Climate-Brief-12-The-EU-ETS-carbon-price-to-intervene-or-not-to-intervene.pdf [Accessed 13.03.2020].

SCHLEICHER, S., MARCU, A., KÖPPL, A., SCHNEIDER, J., ELKERBOUT, M., TÜRK, A. & ZEITLBERGER, A. 2015. *Scanning the options for a structural reform of the EU emissions trading system* [Online]. CEPS (Centre for European Policy Studies). No.107/May 2015. Available:

www.ceps.eu/wp-content/uploads/2015/04/SR107%20Options%20 for%20a%20Structural%20Reform%20of%20EU%20ETS.pdf [Accessed 20.03.2020].

SKJÆRSETH, J. B. & WETTESTAD, J. 2008. Implementing EU emissions trading: Success or failure? *International Environmental Agreements: Politics, Law, Economics*, 8, 275–290.

SKJÆRSETH, J. B. & WETTESTAD, J. 2010. Making the EU Emissions Trading System: The European Commission as an entrepreneurial epistemic leader. *Global Environmental Change*, 20 (2010), 314–321.

STEPHAN, N. & SARTOR, O. 2013. *Reforming the EU ETS: Give it some work!* [Online]. cdc Climate Research. Climate Brief N°28 February 2013. Available: www.i4ce.org/wp-core/wp-content/uploads/2015/09/13-03-06-Climate-Brief-n%C2%B028_Structural-Reform.pdf [Accessed 20.03.2020].

STERK, W., BRAUN, M., HAUG, C., KORYTAROVA, K. & SCHOLTEN, A. 2006b. *Joint emissions trading as a socio-ecological transformation* [Online]. Working Paper I/06. Available: https://wupperinst.org/uploads/tx_wupperinst/JETSET_WP_1-06.pdf [Accessed 02.05.2020].

TUERK, A., MEHLING, M., FLACHSLAND, C. & STERK, W. 2009. Linking carbon markets: Concepts, case studies and pathways. *Climate Policy*, 9, 341–357.

WETTESTAD, J. 2005. The making of the 2003 EU emissions trading directive: An ultra-quick process due to entrepreneurial proficiency? *Global Environmental Politics Climate Law*, 5, 1–23.

WETTESTAD, J. & JEVNAKER, T. 2014. The EU's quest for linked carbon markets. *In:* CHERRY, T. L., HOVI, J. & MCEVOY, D. M. (eds.) *Toward a new climate agreement: Conflict, resolution and governance.* London: Routledge.

4 The EU Emissions Trading System and the California Cap-and-Trade Program

A failed linking attempt

Climate policy in California and the development of the California Cap-and-Trade Program (CAL C&T)

California holds an outstanding position in the USA with respect to its environmental and climate policy making. This reputation dates back at least to the 1970s, when amendments to the Clean Air Act authorized California to adopt stricter automobile emissions standards than those required on a federal level (Vogel, 1997). Experts have claimed that California has the strongest climate policy among the US states (Karapin, 2018).

In addition to the fact that climate change's impacts are already being felt in California, the instability in national climate politics has contributed to California's outstanding role. In the United States, climate and environmental politics are characterized by a back and forth between the federal and the state levels, often depending on the priorities set by the national government and the president (Schreurs, 2010, Wälti, 2009). This situation opened a window of opportunity for US states such as California to establish more ambitious climate policies (Karapin, 2018). Already in the era of the Bush administration, an unintended consequence of his administration's decision to abandon the Kyoto Protocol was the mobilization of subnational climate actors on the state level (Schreurs, 2010). A further driving factor for Californian climate policy is the general support of the public for environmental and climate policy (Baldassare et al., 2017). Surveys show that a majority of Californians believe that climate change is already occurring and the vast majority believe that climate change will occur in the future (California P. P. I. O, 2018). Another aspect that sustains California's role as a frontrunner in climate action is its increasing international activity (California Energy Commission, n.d.). California has become an actor with foreign political relevance even though this domain is formally under the purview of the federal government (Engstrom and Weinstein,

2018). Last but not least, structural features such as economic wealth, a large population and the existence of a strong knowledge economy and intellectual leaders who provide solid scientific expertise on climate change, have been a reason for California's outstanding policy making (Karapin, 2018).

In 2006, California adopted comprehensive climate legislation that included the 'Assembly Bill 32 Global Warming Solutions Act' (AB 32). This law sets overall GHG reduction goals for California: it aims to reduce GHG emissions to 1990 levels by 2020 and to 80% below 1990 levels by 2050 (CARB, 2015). The initial Scoping Plan, released in 2008, defined the launch of an ETS (CARB, 2008). The California Cap-and-Trade Program (CAL C&T) is not the main emissions reductions contributor within the climate policy mix. While the CAL C&T was designed to contribute about 20% (in 2015) to the emissions reductions, other measures, such as the Renewable Energy Portfolio Standard and low carbon fuel standards, are responsible for achieving the remaining 80% (Purdon et al., 2014, Bang et al., 2017). Thus, the CAL C&T works as a backstop mechanism (Gulbrandsen et al., 2019, Bang et al., 2017).

Overall, the adoption of AB 32 and the setting up of an ETS in California built on a multi-year stakeholder process (CARB, 2010a) and was the result of an intense negotiations between affected groups and institutions. When California proposed AB 32, several attempts were made to repeal the legislation. Probably the largest campaign was set against California's central climate policy, AB 32. Fossil fuel industries, such as the coal and petroleum production industries, led the opposition to the climate policy bill and fought against a cap-and-trade system (Karapin, 2018). State referendum 'Proposition 23' sought to suspend AB 32 based on the argument that the bill would increase unemployment and further increase energy costs (Hess, 2014, Elkind et al., 2010). This referendum was challenged by the 'Stop Dirty Energy Committee', a coalition of environmental groups such as the Environmental Defense Fund (EDF), labor associations, social movements such as ethnic minority groups, Governor Arnold Schwarzenegger and the California Air Resources Board (CARB). Many members, such as the EDF, strongly promoted the ETS and helped to design and implement it (EDF, n.d.). (London et al., 2013, Hess, 2014). Proposition 23 was ultimately defeated in 2010, with 62% of votes in opposition to the measure (Purdon et al., 2014: 13).

Nevertheless, the pro-AB 32 coalition was divided when CARB outlined the detailed design of the policy in the Scoping Plan, mandating a cap-and-trade system. Environmental justice groups, primarily, separated themselves from other 'mainstream' environmental groups and public agencies, with the major line of discord being the ETS (London et al., 2013). A principle concern for this movement was the risk of an ETS increasing pollution in low-income or minority communities via the air pollutants that are emitted along with GHGs such as nitrogen oxides, carbon monoxide and particulates (Stavins, 2011). The opposition culminated in 2009, when the environmental justice movement filed a lawsuit before the California Superior Court (London et al., 2013, Carlson, 2011). Even though the launch of the ETS was put on hold in May 2011 by the California Supreme Court, the ruling was overturned by the Court of Appeals in June 2011 and further attempts to stop the system were denied (Walker, 2011). Close to the launch of the CAL C&T, another field of legal disputes emerged on the specific design of the ETS, for example the amount and type of accepted offset credits, and the rules for auctioning them off (EDF and IETA, 2014, Purdon et al., 2014, Murza, n.d., Sabin Center for Climate Change Law, 2013).

Additionally, the establishment of WCI in 2007 played an important role in launching the Californian ETS (Bang et al., 2017). The WCI is a regional non-binding agreement between five North American provinces (California, British Columbia, Manitoba, Ontario, and Québec), which installed a technical forum to helps members get technical and administrative support for climate change mitigation. Originally, the WCI provided design recommendations for the regional ETS, facilitating the harmonization of carbon pricing initiatives of its members.

The CAL C&T was launched in 2012, but started operations with compliance obligations for emissions in 2013 (CARB, 2015). While initially only covering the electricity sector and large-scale manufacturing, the CAL C&T today includes many different emitting sectors ranging from the power sector to industrial processes and transport (CARB, 2015). It covers CO_2, CH_4, N_2O, SF_6, HFC, PFC, NF_3, and other fluorinated gases. The system has an absolute cap that declines every year by roughly 3% (e.g. the cap for 2015 was 394.5; for 2016: 382.4; and for 2017: 370.4 $MtCO_2$e). In 2018 the CAL C&T entered a third compliance period (2018–2020) starting with a cap of 358.3 $MtCO_2$e (ICAP, 2019a). In total, the system covered approximately 85% of California's total GHG emissions as of 2015 (CARB, 2015).

Up to 8% of a facility's compliance obligation can be submitted in offset credits that originate primarily from emissions reduction projects in North America and some Latin American countries (Lake, 2013, CARB, 2015) (CARB, 2010b). Offsets originate from projects in five areas: forestry, urban forestry, dairy digesters, destruction of ozone-depleting substances and mine methane capture (CARB, 2015). Initially, most allowances were allocated freely to participating companies.[1] This practice later transitioned into auctioning the majority of allowances. In 2017, almost 70% of allowances were auctioned (ICAP, 2019a). For industrial facilities, free allocation is decided upon based on the calculation of carbon leakage risk, sector-specific benchmarks, production volumes and a general cap-adjustment factor (ICAP, 2019a).

In California, ETS is also a source of additional financial revenues. California has generated about 11.4 billion EUR (2012–2019) through auctioning allowances under the ETS (ICAP, 2019b).[2] Auctioning proceeds fund further activities under the AB 32 through the Greenhouse Gas Reduction Fund (ICAP, 2019b, CARB, 2019b).[3] Such recipients include, for example, natural resources and waste, transport and sustainable communities, and clean energy and energy efficiency initiatives, as well as disadvantaged communities (e.g. through energy bill payment assistance) (CARB, 2019a). The reuse or recycling of auctioning revenues is a political priority in California and is part of a strategy to gain public support for the ETS (CARB, 2017). Furthermore, a price floor ensures that the price does not fall below a fixed level. The Auction Reserve Price, i.e. the allowance price, below which bids at auction would not be accepted, started in 2012 at circa 9 EUR per allowance (CARB, 2019b) and increases annually by 5% plus inflation (ICAP, 2020a).[4] Furthermore, an Allowance Price Containment Reserve that guarantees a certain maximum price is operated in order to hold costs for companies below a certain limit. In 2017, the CAL C&T was extended until 2030. The bill, AB 398, authorizes the CARB to continue the ETS with some design amendments, such as the reform of the price ceiling provision (ICAP, 2020a).

Linking the EU ETS and the CAL C&T

The linking activities between the EU COM and CARB, the responsible agency in California, reviewed for this book, occurred during 2008–2011. Even though Californian legislation already foresaw the launch of an ETS, the CAL C&T was still in a design phase and not yet officially launched at the time that linking considerations between both partners began. European linking intentions were initially focused on a transatlantic cooperation, as the US was considered an attractive linking partner for the EU (Zetterberg, 2012) (Interview 16). In a 2008 press release the EU stated: "The EU is keen to work with the new US administration to build a transatlantic and indeed global carbon market to act as the motor of a concerted international push to combat climate change" (EU COM, 2008b). However, after several attempts to implement a national US ETS failed, the EU's interest shifted toward

California (Zetterberg, 2012) (Interview 16). In 2009, the EU altered its Linking Directive that formerly prohibited a link to Non-Kyoto Parties and thus enabled linkage to North America, especially California. California also stated its explicit interest in linking its system with the EU ETS. In 2007, government officials were optimistic about the possibility of becoming the first non-European region to connect with the EU ETS (Reuters, 2007). An executive order issued by the former governor Arnold Schwarzenegger appealed for a program that would facilitate linking to the EU ETS and other regions (Mehling, 2007).

At the time of the development of the CAL C&T, exchange and cooperation between European and Californian administrative staff intensified and informal talks between officials on the possibility of linking between both jurisdictions took place (Interviews 14, 15, 16). Nevertheless, the linking activity stagnated in the first phase of the linking process. Although there was no official decision against linking, talks were abandoned before they could reach a more official stage. Yet, the idea of linking has not been completely forgotten. In 2018, California and the EU started to intensify their cooperation again, seeking harmonization as a first step (EU COM, 2018). This can be seen as a signal of renewed interest in that direction. However, this new cooperation does not figure in the time frame of the activity analyzed within this book, which for EU ETS–CAL C&T is 2008–2011.

From a technical perspective, several authors identified the differing use of offsets as a significant obstacle for linking between the EU ETS and the CAL C&T (Zetterberg, 2012, Sterk et al., 2009, Flachsland et al., 2008). The CAL C&T was expected to accept exclusively offsets from projects in North America, and also to include certificates for the absorption of carbon dioxide through the forestry sector (carbon sinks) (ICAP, 2015a). By contrast, the EU excludes projects and credits from the forestry sector but allows for a quota of credits from international CDM projects (ICAP, 2015a). However, the design difference that has been most prominently featured in linking research is the differing price management provisions in California and the EU (Zetterberg, 2012, Carbon Market Watch, 2015b) (Interviews 7, 14, 16, 19). This discrepancy was identified as a possible factor that could preclude a linkage (Haites, 2014, Hawkins and Jegou, 2014, Burtraw et al., 2013, Ranson and Stavins, 2016, Zetterberg, 2012, Flachsland et al., 2008). Specifically, California had decided to introduce a price floor to its ETS at some point during the linking considerations, whereas the EU ETS does not limit the price. Another aspect that would have needed special consideration was the two regions' differing sectoral focuses (Interview 14), as California, unlike the EU, had planned to include the transport sector. Last but not least, if linking had overlapped with the time that California engaged in negotiations with Québec, then that linkage would have had to be taken into account as well. Also, high price fluctuations increased the uncertainty that characterized the linking attempt. After the EUA price collapse in 2007, when linking considerations started in 2008, the EU allowances experienced a brief price peak of about 25–28 EUR, which was much higher than the expected Californian prices. Yet, the EU allowance price then dropped, fluctuating between circa 8–17 EUR after 2009. In 2011, a falling trend, (EUA prices between approximately 5–8 EUR) started to show. Initiated in 2012, the Californian minimum price was set to around 9 EUR.

In the case of linking between the EU ETS and the CAL C&T, the drivers and impediments to the initiative can be attributed to those political parameters; domestic interests, domestic structural elements as well as international developments, described in Chapter 2 of this book.

It can be argued that *economic interests* have acted as a driver and barrier for linking; however, they cannot be considered the single most important factor for this linking attempt. Experts from both jurisdictions argue that the vision of a larger and more economically efficient market motivated linking considerations in California and the EU (Interviews 14, 19, 12) (Pinkster, 2015). Additionally, several academic experts support the view that the carbon price level and economic efficiency would have been important determinants for a link between the EU ETS and CAL C&T (Interviews 3, 6, 7). More importantly, efficiency improvement was used as an argument to 'sell' linking to domestic stakeholders and to obtain consent from industrial groups for linking (Interview 14, 15). Even though both jurisdictions agreed on the possible existence of a general economic benefit to linking, they diverged on the details, such as their visions of ETS price policy and economic priorities. In California, the carbon price range was important, because it expressed a clear vision of what is acceptable for the Californian economy, and what is necessary to protect it (Interview 14). This was reflected in California's decision to impose a clear price floor and ceiling for allowance purchases, in order to ensure emissions reductions while protecting the economy from unbearably high costs (Interviews 14, 19). In the EU, the introduction of a minimum price in order to make the EU ETS more compatible with the CAL C&T would have been unrealistic (Interview 16). Furthermore, it can be assumed that due to the perceived growing risk for low carbon prices in the EU, a potential flow of funds from California to the EU became a reason for not pursuing the link between California and the EU (Interviews 16, 6,

7). Had linking talks become more concrete, such economic interests would likely have been important to consider (Interviews 4, 5, 6, 7).

During the analyzed time period, the principal *environmental* and *climate policy-related interest* among the involved stakeholders was to establish of a stringent, well-functioning system to achieve climate targets (Interviews 14, 19, 16, 12, 15). At the same time, developments in the EU—namely the price collapse of 2009 in the EU ETS and a visible trend toward of a structurally oversupplied market—had negative consequences for the linking attempt. This situation decreased the EU's attractiveness as linking partner (Interviews 15, 16). This is evidenced by the fact that some stakeholders in California used the EU ETS price collapse as an argument against linking with the EU; in doing so, they highlighted suspicions that the EU ETS was not working environmentally effectively, had a weak design, and would ultimately negatively influence the CAL C&T (Interviews 14). Moreover, while Californian policy-makers have not claimed explicitly that such assumed lack in stringency ended their interest in linking, they have noted that the ability of the linking partners to demonstrate environmental stringency and integrity is crucial, and that the technical challenge around the oversupply situation in the EU was being observed very closely (Interviews 19, 14). Likely, California's hesitancy was more due to the uncertainty of how this situation would be handled in the EU and the fear of potential stakeholder concerns. Linking to such a 'moving target' was perceived as even more risky, when the CAL C&T had not yet been launched, and the impact of EU ETS developments were difficult to estimate. Furthermore, in California, the achievement of reductions *within* the jurisdiction was politically demanded through pressure by several interest groups and the legislature (Interviews 14, 19). California's preference for domestic emissions reduction comprises a trade-off within the broader idea of international linking, and differed from the goals of the EU ETS, where no such priorities existed.

The conceptual framework underlying this book postulate that linking could affect *co-benefits* expected from ETS policy. In California, the raising and spending of ETS proceeds has been a politically sensitive issue (Interview 14), because it has acted as an argument to convince the constituents of the value of the system (Interview 8). The link to a lower priced EU ETS would have borne the risk of lower financial income for California. However, in the observed linking activity between the EU and California, this was not observed as a central influencing factor, likely, because the ETS had not started to operate and such financial co-benefits were yet to be experienced.

Data collected for this research suggests that aspirations for being a *role model and leader* both drove, but then also inhibited, the linking initiative. An EU link with California would have had a symbolic meaning, sending a clear signal for international collaboration (Interviews 16, 11). The EU had an interest in showing not only that its ETS worked well and could attract followers, but also in demonstrating that linking was feasible (Interviews 12, 15, 11). Such a link would have meant a general upgrade of the status of ETS as a policy instrument, by demonstrating that ETS was not simply an 'EU only' idea, but one that was becoming globally rooted and leading to collaboration (Interviews 12, 11). Some experts suggest that the EU implemented its ETS in the early years with the idea of becoming the largest and most important ETS globally, one that many other countries would like to be connected to (Interview 2). In doing so, it was believed that the EU would become a climate policy frontrunner with its ETS serving as a role model (Interviews 6, 11, 12). But also for California, building on its established role as a climate policy frontrunner, being a leader and role model has been very important (Interviews 6, 8). Policy experts describe this as a main motivation for the launch the CAL C&T as well as the intent to design it in an ambitious and effective form that could be copied adapted and improved on (Interviews 14). In the beginning, there was the expectation of becoming a role model for the US national level, placing California in the position to influence future national ETS rules (Interviews 19, 6). It appears that both jurisdictions had set their sights on being 'rule makers' and role models, and this created a difficult environment for negotiations. Additionally, while linking to another strong and stringent ETS could, in principle, foster the attractiveness of an ETS, in this particular case Californian policy-makers likely had not expected reputational gains, because of the EU ETS' allowance over-supply problems.

Leadership and being a role model position can put a region in *a strong and powerful negotiation position*, allowing them to set the standards rather than adopting them. Even though the CAL C&T was not expected to cover an equal amount of GHGs, nor was California seen as equally economically powerful to the EU, the perceived imbalance of power was rather small. EU policy-makers saw the CAL C&T on a rather equal standing with the EU ETS (Interviews 16, 8). EU officials saw California as a strong and, in many ways, powerful, linking counterpart that would not simply accept being adjoined to the EU ETS (Interview 2). Somehow, from California's perspective, the power balance was seen as more mixed. The larger size of the EU ETS would have caused an imbalance of influence from a market perspective, something that was seen as a more diffuse risk. Policy-makers mentioned this market imbalance as something that would have increased the tensions in California and therefore made the potential loss of control over the domestic ETS policy much more prevailing and complex (Interview 14).

Likely the most palpable difference between the EU and California, their *legal status,* has been recognized widely by academic and policy experts as a barrier to their linking attempt, as it added complexity and uncertainty to the process (Mehling, 2007, Zetterberg, 2012) (Interviews, 1, 2, 5, 7, 12, 15, 14, 16, 19). However, this analysis found a mixed picture of what specifically played a role and for whom. The US constitution limits the foreign political capacities of states, which also applies to international climate policy. According to Article 1, Section 10 of the US Constitution, foreign policy is national domain and states are not allowed to enter into any formal treaty with other countries (Library of Congress, n.d.). However, states may adopt a binding 'compact' or 'agreement' with foreign powers, such as MoUs, which requires congressional consent only if the compact increases the political power of the state (Mehling, 2007). Thus, uncertainty figured into the question of whether California could link to the EU and on what basis (Interviews 2, 7, 16). Californian policy-makers were uncertain about if and how they would have to involve the federal government of the US in such negotiations. Despite their above noted growing international activity, they maintained a very careful approach to linking with the EU (Interviews 14, 19). Also, EU policy-makers hesitated to formulate talks more concretely out of fear of instigating conflicts or unanimities with the US federal government and the resulting disadvantages this might have brought with it (Interviews 15, 16). In other words, independent action on linking could have been regarded as an affront by the US federal level. It would have needed a federal 'green light' (Interviews 6, 12). Furthermore, as a subnational region, California was faced with the challenge that the federal level could impose a policy that takes priority over state law, thereby also affecting the ETS. US federal legislation to regulate GHGs, such as the 2014-initiated Clean Power Plan (CPP) (US EPA, 2014), could preclude state law (Mehling, 2007).[5] Altogether, there was an overall expectation on California's end that national policy making would affect linking, and this led to some hesitation about broadening the system internationally (Interview 1).

Last but not least, the fact that linking with the EU was not a top priority on the *political agenda* for CARB had a rather negative influence on the progress of this linking attempt (Interview 14). During the examined time periods, policy-makers from both systems were still concentrating on making their ETSs work properly. The Californian government was also busy with the stakeholder conflicts initially described in this chapter and pushing back at attempts to repeal the ETS (Interview 8). Meanwhile in the EU, EU COM was occupied with its own technical challenges, such as the oversupply and cyber frauds (Interview 15). EU experts named improving the domestic system as a strong argument for their hesitancy toward linking (Interviews 15, 16). The preparation of a structural reform (which was proposed in 2012) took significant capacity away from a linking assessment, and clearly took priority in the EU's climate policy and ETS agenda (Interviews 15, 1).

As linking efforts were kept very informal between government officials, *involvement of European* and *Californian stakeholders* was apparently limited in this case. In California, influential industrial constituents with obligations under the EU ETS, such as international oil and electricity companies, foresaw advantages that would come from linking to a larger market and therefore generally supported linking (Interview 14). This group promoted the possibility of linking to the EU ETS, because they saw it as an opportunity to cope more effectively with both programs (Interview 14). In contrast, several environmental groups raised concerns about linking, labelling it as a risk for the CAL C&T's environmental stringency and integrity (Gulbrandsen et al., 2019, Bang et al., 2017). Critical voices became loud when the EU ETS faced the oversupply situation and prices collapsed, and this was used as an argument against linking to the EU (Interview 14). All in all, it is likely that the stakeholder conflicts and legal disputes over ETS policy in general that California was confronting at the time acted as limiting factors to linking, because policy-makers expected that linking to the EU would increase these tensions (Interview 14). For the EU, this book's research noted very little stakeholder advocacy or opposition on the linking initiative with California. However, European policy-makers tend to be very sensitive to the potential reactions of stakeholders (Interviews 16, 15). Thus, had the link become more concrete, maybe more tensions could have been expected. Especially in the first years of the EU ETS, European businesses were enthusiastic about linking, with some groups even demanding links to other ETSs worldwide as a policy recommendation to improve cost efficiency (Business Europe, 2013) (Interview 5). Still, EU businesses' perceived critical differences between the EU and California price policies; the Californian price floor was generally viewed with concern, but the price ceiling as an interesting option (Interview 5). However, neither the potential gains nor hurdles mobilized European industrial stakeholders to take strong action.

Research carried out for this book has not found evidence that the notion of *political, geographical,* and *cultural proximity* has had a significant role in the linking attempts between the EU and California. However, it is likely that existing transatlantic relationships and cultural closeness, in, for example, the English language, created a somewhat beneficial environment for linking. Also, the technical exchange facilitated through mutual participation in international fora such as ICAP and IETA probably provided a common ground for linking talks.

International and jurisdiction-external developments during 2008–2011 have clearly had an impact on the linking activity between California and the EU. For California, the absence of national climate regulation became a reason to pursue a strong, frontrunner role through the CAL C&T (Interviews 1, 14). In the EU the reason for turning to California was that at some point, it became clear that no US national ETS would be launched (Interviews 7, 15). Furthermore, the rise of new actors, especially the announcement of a Chinese national ETS, was described as a trigger for the shift of the EU's attention from the United States to the East, and especially China (Interviews 15, 16, 11). At the same time, other subnational systems like Québec's ETS were emerging in North America and began to receive California's attention. The QUE C&T is a much smaller system with an almost identical design to the CAL C&T, so it was therefore perceived as the easier linking partner (Interview 14). While this analysis cannot pinpoint exactly when linking to the QUE C&T withdrew California's attention from the EU as a potential linking partner, this development shows that it was easier for California to concentrate on North America, a region where California had already been a leader. Last but not least, also in relation to the process under the UNFCCC, another structural challenge for linking between the CAL C&T and the EU ETS came into play: The US as a whole, contrary to the EU, is not party to the Kyoto Protocol. California, as a subnational entity, could not join this international agreement on its own. Thus, the CAL C&T could not issue emissions allowances equivalent to units dealt under the Kyoto Protocol, since these were reserved to Kyoto parties. Therefore, from both a legal and political perspective, the EU could not have used Californian emissions allowances under its Kyoto commitments and this limitation was seen as problematic for linking (Interviews 16, 6). Because the EU ETS guaranteed the largest part of the EU's emissions reduction commitments under the Kyoto Protocol, this challenge might have required a different accounting system and, theoretically, additional political efforts from the EU.

Notes

1 In California, a distinction is made between state-owned allowances and consigned allowances to utilities.
2 12.5 billion USD.

3 All foreign currencies (USD) were converted by the author, as of 03.05.2020 (https://bankenverband.de/service/waehrungsrechner/).
4 10 USD.
5 In 2014, the United States initiated the CPP to regulate GHGs on a federal level, but the policy had to be implemented through the states, e.g. through ETS polices. As the regulation only covered the domestic electricity sector, it prohibited counting sources of reductions from additional offset credits or credits of other international ETSs against their obligations within the electricity sector (US EPA, 2014). This would have complicated any linkage to the existing ETSs worldwide, apart from RGGI, including its link to the QUE C&T.

References

BALDASSARE, M., BONNER, D., KORDUS, D. & LOPES, L. 2017. *Californians & the environment* [Online]. Available: www.ppic.org/wp-content/uploads/s_717mbs.pdf [Accessed 24.10.2017].

BANG, G., VICTOR, D. G. & ANDRESEN, S. 2017. California's cap-and-trade system: Diffusion and lessons. *Global Environmental Politics*, 17, 12–30.

BURTRAW, D., PALMER, K. L., MUNNINGS, C., WEBER, P. & WOERMAN, M. 2013. *Linking by degrees: Incremental alignment of cap-and-trade markets* [Online]. Resources for the Future 13-04. Available: www.rff.org/publications/working-papers/linking-by-degrees-incremental-alignment-of-cap-and-trade-markets/ [Accessed 12.03.2017].

BUSINESS EUROPE. 2013. A Competitive EU Energy and Climate Policy [Online]. Available: https://english.bdi.eu/media/topics/europe/publications/201306__Brocure_A_competitive_EU_energy_and_climate_policy.pdf [Accessed 02.04.2020].

CALIFORNIA ENERGY COMMISSION. n.d. *Climate change partnerships* [Online]. Available: www.energy.ca.gov/about/campaigns/international-cooperation/climate-change-partnerships [Accessed 03.04.2020].

CALIFORNIA, P. P. I. O. 2018. *Californians' views on climate change* [Online]. Available: www.ppic.org/publication/californians-views-on-climate-change/ [Accessed 10.11.2018].

CARB. 2008. *Climate change scoping plan* [Online]. Available: https://ww3.arb.ca.gov/cc/scopingplan/document/adopted_scoping_plan.pdf [Accessed 02.03.2017].

CARB. 2010a. *Staff report: Initial statement of reasons proposed regulation to implement the California Cap-And-Trade program volume II appendix D public hearing to consider the proposed regulation to implement the California Cap-and-Trade program* [Online]. Available: https://ww3.arb.ca.gov/regact/2010/capandtrade10/capv2appd.pdf [Accessed 02.04.2020].

CARB. 2010b. *Memorandum of understanding on environmental cooperation between the state of acre of the Federative Republic of Brazil, the State of Chiapas of the United Mexican States and the State of California of the United States of America* [Online]. Available: https://ww3.arb.ca.gov/cc/capandtrade/sectorbasedoffsets/2010%20mou%20acre-california-chiapas.pdf [Accessed 02.04.2020].

CARB. 2015. *Overview of ARB emissions trading program* [Online]. Available: www.arb.ca.gov/cc/capandtrade/guidance/cap_trade_overview.pdf [Accessed 23.3.2020].

CARB. 2017. *Cap-and-Trade benefits all Californians. Scoping plan factsheet* [Online]. Available: https://ww3.arb.ca.gov/cc/scopingplan/2017sp_factsheet.pdf [Accessed 02.04.2020].

CARB. 2019a. *Annual report 2019. Cap-and-Trade auction proceeds.* [Online]. Available: https://ww3.arb.ca.gov/cc/capandtrade/auctionproceeds/2019_cci_annual_report.pdf [Accessed 02.04.2020].

CARB. 2019b. *Unofficial electronic version of the Regulation for the California Cap on greenhouse gas emissions and market-based compliance mechanisms* [Online]. Available: https://ww3.arb.ca.gov/cc/capandtrade/ct_reg_unofficial.pdf [Accessed 02.04.2020].

CARBON MARKET WATCH. 2015b. *Towards a global carbon market. Risks of linking the EU ETS to other carbon markets* [Online]. Available: https://carbonmarketwatch.org/wp-content/uploads/2015/05/NC-Towards-a-global-carbon-market-PB_web.pdf [Accessed 04.04.2020].

CARLSON, A. 2011. *AB32 lawsuit: Assessing the environmental justice arguments against Cap and Trade* [Online]. Berkeley Law, UCA Law. Available: https://legal-planet.org/2011/03/22/ab-32-lawsuit-assessing-the-environmental-justice-arguments-against-cap-and-trade/ [Accessed 15.08.2017].

EDF. n.d. *How cap and trade works* [Online]. Available: www.edf.org/climate/how-cap-and-trade-works [Accessed 02.05.2020].

EDF & IETA. 2014. *California, the world's carbon markets: A case study guide to emissions trading* [Online]. Available: www.edf.org/sites/default/files/California-ETS-Case-Study-March-2014.pdf [Accessed 14.04.2017].

ELKIND, E., FARBER, D., FRANK, R., HANEMANN, M., KAMMEN, D., KANTENBACHER, A. & WEISSMAN, S. 2010. *California at the crossroads: Proposition 23, AB 32, and climate change* [Online]. Center for Law, Energy the Environment, Berkeley Law at the University of California, Berkeley. Available: www.law.berkeley.edu/center-article/california-at-the-crossroads-proposition-23-ab-32-and-climate-change/ [Accessed 02.03.2020].

ENGSTROM, D. F. & WEINSTEIN, J. M. 2018. What if California had a foreign policy? The new frontier of states' rights. *The Washington Quarterly*, 41, 27–43.

EU COM. 2008b. *Questions and answers on the revised EU emissions trading system* [Online]. Available: https://ec.europa.eu/commission/presscorner/detail/en/MEMO_08_796 [Accessed 02.05.2020].

EU COM. 2018. *EU and California to step up cooperation on carbon markets* [Online]. Available: https://ec.europa.eu/clima/news/eu-and-california-step-cooperation-carbon-markets_en [Accessed 15.12.2018].

FLACHSLAND, C., O. EDENHOFER, M. JAKOB & STECKEL, J. 2008. *Developing the international carbon market. Linking options for the EU ETS* [Online]. Report to the Policy Planning Staff at the German Federal Foreign Office. Available: www.mcc-berlin.net/fileadmin/data/pdf/PIK_Carbon_Market_Linking_2008.pdf [Accessed 03.02.2019].

GULBRANDSEN, L. H., WETTESTAD, J., VICTOR, D. G. & UNDERDAL, A. 2019. The political roots of divergence in carbon market design: Implications for linking. *Climate Policy*, 19, 427–438.

HAITES, E. 2014. *Lessons learned from linking emissions trading systems: General principles and applications* [Online]. Washington, DC: Partnership for Market Readiness (PMR). Available: www.thepmr.org/system/files/documents/PMR%20Technical%20Note%207.pdf [Accessed 09.09.2019].

HAWKINS, S. & JEGOU, I. 2014. *Linking emissions trading schemes: Considerations and recommendations for a joint EU-Korean carbon market* [Online]. ICTSD Global Platform on Climate Change, Trade and Sustainable Energy Available: www.ictsd.org/sites/default/files/research/linking-emissions-trading-schemes-considerations-and-recommendations-for-a-joint-eu-korean-carbon-market.pdf [Accessed 09.09.2019].

HESS, D. J. 2014. Sustainability transitions: A political coalition perspective. *Research Policy*, 43, 278–283.

ICAP. 2015a. *Emissions trading worldwide. International carbon action partnership status report 2014* [Online]. Available: https://icapcarbonaction.com/en/?option=com_attach&task=download&id=349 [Accessed 13.08.2017].

ICAP. 2019a. *Emissions trading worldwide international carbon action partnership (ICAP) status report 2018* [Online]. Berlin. Available: https://icapcarbonaction.com/en/?option=com_attach&task=download&id=625 [Accessed 23.05.2019].

ICAP. 2019b. *From carbon market to climate finance: Emissions trading revenue* [Online]. ETS BRIEF #5. Available: https://icapcarbonaction.com/en/?option=com_attach&task=download&id=674 [Accessed 16.03.2020].

ICAP. 2020a. *Emissions trading worldwide: International carbon action partnership (ICAP) status report 2019* [Online]. Berlin. Available: https://icapcarbonaction.com/en/?option=com_attach&task=download&id=677 [Accessed 02.05.2020].

KARAPIN, R. 2018. Not waiting for Washington: Climate policy adoption in California and New York. *Political Science Quarterly*, 133, 317–353.

LAKE, K. 2013. *Comment: Linking Australia's carbon trading to Europe's ETS* [Online]. Available: SBSNewspostComment_ Linking Australia's carbon trading to EU ETS_2013.pd [Accessed 05.12.2014].

LIBRARY OF CONGRESS. n.d. *Constitution of the United States* [Online]. Available: https://constitution.congress.gov/constitution/ [Accessed 03.04.2020].

LONDON, J., KARNER, A., SZE, J., ROWAN, D., GAMBIRAZZIO, G. & NIEMEIER, D. 2013. Racing climate change: Collaboration and conflict in California's global climate change policy arena. *Global Environmental Change*, 23, 791–799.

MEHLING, M. 2007. Bridging the transatlantic divide: Legal aspects of a link between regional carbon markets in Europe and the United States. *Sustainable Development Law & Policy*, 7, 46–51.

MURZA, M. n.d. *Recommendations for a national Cap-and-Trade system based on the successes and failures of the three largest existing carbon markets*

[Online]. Available: https://law.ucdavis.edu/centers/environmental/files/Recommendations-for-a-National-Cap-and-Trade-System.pdf [Accessed 02.05.2020].

PINKSTER, J. 2015. *EU perspective on linking* [Online]. Ministry of Infrastructure and Environment. Available: http://iki-alliance.mx/wp-content/uploads/2015-01-29_DAY4_Presentation-Pinkster_Linking-in-the-European-ETS.pdf [Accessed 04.04.2020].

PURDON, M., HOULE, D. & LACHAPELLE, E. 2014. *The political economy of California and Québec's Cap-and-Trade systems* [Online]. Available: https://institute.smartprosperity.ca/sites/default/files/publications/files/QuebecCalifornia%20FINAL.pdf [Accessed 17.04.2017 Sustainable Prosperity Research Report].

RANSON, M. & STAVINS, R. N. 2016. Linkage of greenhouse gas emissions trading systems: Learning from experience. *Climate Policy*, 16, 284–300.

REUTERS. 2007. *California eyes joining EU emissions trading scheme* [Online]. Available: www.reuters.com/news/picture/california-eyes-joining-eu-emissions-tra-idUSL2955639620070329 [Accessed 12.03.2015].

SABIN CENTER FOR CLIMATE CHANGE LAW, C. U. 2013. *California Chamber of Commerce v. California Air Resources Board* [Online]. Available: https://climatecasechart.com/case/california-chamber-of-commerce-v-california-air-resources-board/ [Accessed 02.04.2020].

SCHREURS, M. A. 2010. Climate change politics in the United States: Melting of the ice. *Analyse & Kritik*, 2010, 177–189.

STAVINS, R. 2011. *Why the environmental justice lawsuit against California's climate law is misguided* [Online]. Grist Magazine, May 23, 2011. Available: https://grist.org/climate-policy/2011-05-23-environmental-justice-lawsuit-against-californias-climate-law/ [Accessed 17.09.2019].

STERK, W., MEHLING, M. & TUERK, A. 2009. *Prospects of linking EU and US emission trading schemes: Comparing the Western Climate Initiative, the Waxman-Markey and the Lieberman-Warner Proposals* [Online]. Climate Strategies. Available: https://climatestrategies.org/wp-content/uploads/2009/04/linking-eu-us.pdf [Accessed].

US EPA. 2014. *Fact sheet: Clean power plan technical support document.* [Online]. Available: https://archive.epa.gov/epa/cleanpowerplan/fact-sheet-clean-power-plan-technical-support-document.html [Accessed 04.05.2018].

VOGEL, D. 1997. Trading up and governing across: Transnational governance and environmental protection. *Journal of European Public Policy*, 4, 556–571.

WALKER, L. Z. 2011. *Cap and Trade regulations approved and transmitted; Preliminary list of covered entities now available* [Online]. Land Use Law Blog. Abbott & Kindermann, Inc. Available: https://blog.aklandlaw.com/2011/12/articles/climate-change/cap-and-trade-regulations-approved-and-transmitted-preliminary-list-of-covered-entities-now-available/ [Accessed 03.06.2015].

WÄLTI, S. 2009. Intergovernmental management of environmental policy in the United States and the EU. *In:* SCHREURS, M. A., SELIN, H. &

VANDEVEER, S. D. (eds.) *Transatlantic environment and energy politics: Comparative and international perspectives.* Farnham: Ashgate.
ZETTERBERG, L. 2012. *Linking the emissions trading systems in EU and California* [Online]. Fores, Swedish Environmental Research Institute. Available: https://fores.se/wp-content/uploads/2013/04/FORES-California_ETS-web.pdf [Accessed 05.05.2019].

5 A successful linking process between the EU Emissions Trading System and the Switzerland Emissions Trading System

Swiss climate policy and the implementation of the SW ETS

On a global scale, Switzerland is a relatively small emitter of GHGs, with 47.24 Mt CO_2e in 2017 (BAFU, 2019a). The transport, building, and industry sectors account for the largest share of GHGs with, respectively, 32%, 27%, and 23% in 2017 (BAFU, 2019a). The central climate legislation, the CO_2 Act, was adopted in 2011 and focuses on reducing Switzerland's GHG emissions. It commits to a 20% domestic reduction from 1990 GHG levels by 2020 (The Federal Council Government of Switzerland, 2011). The CO_2 Act concentrates on the three major emitting sectors—transport, buildings, and industry—and covers a diverse range of policy instruments. The central instrument is the CO_2 levy, which was imposed in 2008 on fossil fuels such as oil and natural gas. As of 2018, it amounts to 91 EUR[1] per emitted ton of CO_2 (BAFU, 2019d).[2] Fossil fuel importers are required to compensate their emissions through domestic emissions reductions projects (BAFU, 2019b). In addition, an ETS was launched in 2008 to tackle the industrial sector and to contribute to Switzerland's commitments under the Kyoto Protocol (BAFU, 2018). Under the Paris Agreement, Switzerland plans to achieve a 50% reduction from 1990 GHG levels by 2030. This target comprehends both domestic reductions as well as reductions achievable through international offsets. In order to achieve the objectives committed under the Paris Agreement, Switzerland started a revision process of the CO_2 Act in 2017, which is still not concluded as of writing (Bundesversammlung Schweiz, 2020).

Switzerland's climate legislation, policy instruments, and targets have to be regarded as a compromise solution resulting from a long negotiation process between all interest groups (Oberauner and Krysiak, 2008). Climate policy in Switzerland has been challenged by its own set of conflicts, for example, in response to the 2009 emissions reduction targets discussions (Groß, 2011). When the complete revision of the CO_2 Act (Switzerland, 2017) failed in December 2018, this put the achievement Switzerland's GHG reduction targets promised under the Paris Agreement at risk (Institute and Analytics, 2019). Furthermore, a glimpse into Swiss media coverage of the ETS implementation reveals discussions over the economic efficiency and environmental effectiveness of the ETS (Neue Zürcher Zeitung, 2009). As the motivation for establishing an ETS was, in large part, to alleviate the financial burden of industry (Interview 10), it was mostly welcomed by industrial groups. Only very few positions of environmental or societal groups on the implementation of ETS as climate policy tool can be found. A press article suggests, for example, that the World Wide Fund for Nature (WWF) Schweiz was generally open to an ETS in Switzerland, but that it demanded a more ambitious GHG reduction target for 2020 (Neue Zürcher Zeitung, 2009).

The SW ETS was conceptualized as an alternative to the Swiss CO_2 levy. The EU ETS has functioned as a blueprint for the Swiss ETS (Interview 13). In 2008, the ETS began with a five-year pilot phase, where participants could join voluntarily. Participation became mandatory for the second phase (2013–2020), with compliance obligations starting first for large, energy-intensive industries. Companies have the possibility to 'opt out' of the system, if their total emissions have resulted in less than 25,000 CO_2e over the last three years. Medium size emitters have the option to join-in and voluntarily participate in the system, in order to be exempted from the CO_2 levy (BAFU, 2019b).

In its current state, the SW ETS covers companies from the cement, chemistry, pharma, refineries, paper, district heating, steel and other sectors (BAFU, 2019b). As of 2016, the ETS covered about 10% of the total Swiss GHGs and covers approximately 55 companies.

The overall cap was set at 5.63 $MtCO_2$ in 2013 and subsequently declined under the linear reduction factor of 1.74 (of the 2010 total amount) per annum, until reaching 4.9 Mt CO_2 in 2020 (BAFU, 2019b). While in the first phase all allowances were handed out for free, Switzerland gradually started to auction a portion of certificates during the mandatory phrase. Free allocation is distributed according to benchmarks and includes a calculated carbon leakage factor (BAFU, 2019b). The SW ETS allows for the use of a certain quota of offsets, for example, those originating from CDM projects and other 'Kyoto units': ERUs, CERs, and Removal Units (RMUs). Certificates from carbon sink projects such as afforestation and reforestation were initially allowed but could not be banked for future commitment periods (EDF and CDC Climat Research, 2015, Carbon Market Watch, 2015a). In case a company fails to comply with the obligation, it is forced to pay a penalty of 118.4 EUR/tCO_2 and make up for the missing allowances in the following year.[3]

Linking the EU ETS and the SW ETS

It is difficult to recapture the exact moment when Switzerland and the EU started to discuss linking their ETS, as unofficial talks were already being conducted in 2004 (Interview 10, 11) before either system had started to operate. The Swiss responsible authority, the Bundesamt für Umwelt (BAFU), officially reports to have started explorative talks in 2008 (BAFU, 2009) and in December 2009, the Swiss legislature provided the mandate to start bilateral linking negotiations with the EU (BAFU, 2009). The EU COM officially requested a mandate from the EU COU in November 2010 (EU COM, 2010). Five rounds of negotiations took place between March 2011–December 2013 (BAFU, 2013, BAFU, 2011). The initial vision of a linked market by 2011 was changed to a goal of completing negotiations in spring 2014 (Oberauner and Krysiak, 2008). However, in spring 2014, a rupture in the bilateral relationship brought negotiations to an immediate pause. Independent of the ETS policy, in a public referendum Swiss voters decided that they wanted to restrict immigration to Switzerland, including the free movement of European citizens to Swiss territory (EU PA, 2014, European Voice, 2014). As a consequence, the EU stopped the negotiations on linking the ETSs. Although negotiations were reopened later in 2014, the conflict continued in the background. When negotiations on the ETS linkage were officially concluded in January 2016 and an agreement was drafted, the EU stated that the signing of the treaty depended on solving the immigration issue (BAFU, 2016). The Swiss Federal Council approved the signing of the linking agreement in August 2017, after new legislation that regulated the migration issue was introduced and the Council of the European Union authorized the signing of the linking treaty in November 2017 (BAFU, 2019c). At the end of 2019, the agreement was ratified by both partners and their respective legislations were amended. After this step, the common market came into force as of January 1, 2020 (EU COM, 2019a, EU COM, 2019b, BAFU, 2019c). From this year on, the emission allowances eligible for compliance in one system will be eligible for compliance in the other as well (EU COM, 2019c).

The SW ETS was designed very similarly to the EU ETS. Therefore, the official EU linking conditions for technical comparability, as per Directive, Article 25(1a), were fulfilled (Council of the European Union, 2019). Initially, when linking was first being negotiated in 2008, the SW ETS was voluntary in nature (Interview 10). This was perceived as a technical challenge (Oberauner and Krysiak, 2008). In

2013, in its second phase, the SW ETS was transformed into a mandatory system for those companies not regulated under the CO_2 levy. One technical aspect on which the systems initially diverged, was the use of offsets. To rectify this discrepancy, the SW ETS was altered in 2013 to no longer allow for the use of credits originating from nuclear or LULUCF projects (Rutherford, 2014). The use of offsets from carbon sinks was also met with some conflict, and the rules in the SW ETS were subsequently altered so that Swiss companies would not be allowed bring these credits over to future commitment periods (Carbon Market Watch, 2015a). For linking, the aspect that was perceived as most challenging was the aviation sector, which originally had not been part of the SW ETS. This technical challenge was solved when Switzerland agreed on including the aviation sector into the ETS by the time the link became operational (Interviews 10, 11, 13). Last but not least, further technical-legal aspects, such as the EU's legal definition of the emissions allowances complicated the negotiations (Interviews 13, 15), but the situation was solved when the EU decided to include EUAs as a financial instrument under the Directive on markets in financial instruments issued by the European Parliament and the Council (MiFID II) (Glock et al., 2016). Finally, for the link to function, the registries had to be harmonized in order to allow for the acceptance of allowances from both systems (BAFU, 2016, Santikarn et al., 2018).

At the start of linking and during the course of the negotiations, EUAs were sold at a lower price than Swiss allowances. This situation changed when in 2019 the EUA prices increased after the start of the MSR and fluctuated at around 24.9 EUR (ICAP, 2020c), while the Swiss average auction price was only at about 12 EUR (ICAP, 2020b) (Figure 5.1).[4]

Overall, linking has taken much more time than anticipated. Switzerland and the EU passed through all phases of linking, starting with informal talks and finalizing with a fully functional linked market (Council of the European Union, 2019, EU COM, 2019c). Over the course of the linking activity, several of the parameters considered in this analysis acted to both motivate and deter the link.

Economic interests have clearly driven linking between Switzerland and the EU from a very early stage. Before starting the negotiations, Switzerland commissioned a study that evaluated the potential effects of linking to the EU ETS, which estimated an overall positive economic benefit (ECOPLAN, 2008). For Switzerland, economic interests, through efficiency gains and likely cost savings, have been the main driver for pursuing a link with the EU (Interview 13, 10). Economic benefits expected through the access to a larger and more liquid market

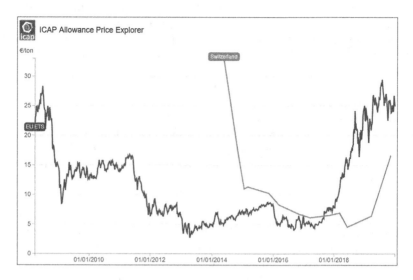

Figure 5.1 Carbon price development in the EU ETS and the SW ETS
          (2008–2019).

Source: ICAP, 2020: Allowance Price Explorer. Retrieved 28.07.2020 from:
https://icapcarbonaction.com/en/ets-prices (used with permission granted by
the ICAP Secretariat)

were also communicated as the greatest advantage of this link to the
Swiss stakeholders and public (BAFU, 2016, BAFU, 2013, BAFU,
2011, BAFU, 2009). The initially cheaper EUAs, and the expectation
of reduced compliance costs post linking, were seen as a benefit for
Switzerland's energy-intensive industry (BAFU, 2009). An additional
advantage was the possibility for companies with businesses in both
jurisdictions to experience a reduction in administrative costs if the
same allowances could be used in both systems. As the SW ETS is very
small (ca. 55 facilities) compared to the EU ETS (ca. 11,000 facilities),
it faced liquidity challenges, which restrained trading, and linking to
the EU ETS promised to solve this problem (Interviews 10, 13) (BAFU,
2011, BAFU, 2013, BAFU, 2016). Conversely, economic benefits have
not been at the forefront of the EU's considerations, because the SW
ETS is very small and policy-makers therefore expected it to have very
little influence on the EU ETS (Interview 15).

  This analysis has found that *climate* and *environmental policy-related
interests* have played only a minor role in the negotiations. In this case,
the growing oversupply and the price collapse in the EU ETS was not

seen as very influential in the negotiations (Interview 13) even though related concerns were occasionally raised by Swiss stakeholder groups. Policy-makers did not perceive the allowance oversupply in the EU ETS as a reason to stop negotiations, in part because after the official start of negotiations in 2011, EU reform efforts were visible (Interview 13). One aspect that might have functioned indirectly as an additional driving force for linking was the position of the SW ETS in the policy mix and its relationship to the CO_2 levy, which generates the largest share of reductions (Interview 13). Companies that pay the CO_2 levy are not covered under the link. While the limited scope of the SW ETS does not mean the SW ETS is an unimportant instrument in Switzerland (Interview 10), this arrangement may have helped facilitate acceptance of the EU's linking conditions.

Also, the research carried out for this book could not find evidence that the expectation of gains or losses of *ETS co-benefits,* such as adherent goals, financial auctioning revenues, or energy prices, played a significant role in the linking negotiations between the EU and Switzerland.

Research has shown that also in this case, the EU displayed an *interest in being a role model and ETS leader.* This aspect has mostly been important to motivate EU policy-makers to link despite there being no large economic gains expected from the link, because linking to the SW ETS demonstrates the success of the EU ETS model. Serving as a role model that incentivizes other regions to create similarly designed ETSs with whom the EU could then link was an important motivation for the EU (Interview 15, 16). By this logic, linking with Switzerland was seen as a way to set an example for how the EU deals with small partners (Interview 16); it would also serve as an example for the European periphery, hopefully leading them to adopt the European approach to ETS (Interview 2, 15, 16). In general, linking to Switzerland as a country in the geographical center of Europe was regarded as a logical, but also easy step (Interviews 15, 16). Yet, the failure of such a link would have been negatively regarded internationally, signaling incapacity on the part of the EU (Interview 15). The longer the negotiations lasted, the more the pressure grew to demonstrate that the EU could effectively realize linking (Interview 11). The need to show that linking is realizable and to display a linking success became more even more urgent after the failure of linking to Australia in 2014 (Interview 16). The link might end up giving Swiss climate policy more international visibility (Interview 10). However, the symbolic value was not the main motivation for Switzerland to link with the EU; such benefits alone would not have been sufficient to convince Swiss constituents of the usefulness of the link (Interview 13).

Likely the most obvious aspects to consider in this linking process are the different *sizes of the EU ETS* and *the SW ETS* and the inherent imbalanced *power constellations* related to economic and political capacity. Switzerland's willingness to accept the EU's dominance has acted as a major driving factor behind linking. However, further aspects of the Swiss–EU relationship have shown that politics and political sensitivities originating from the power imbalance have caused frictions. The EU held a powerful position in these negotiations, because it has the larger carbon market and thereby greater economic benefits to offer. As could be seen with the leadership dimension noted above, the EU also had an interest in holding a position of power and control in order to spread its ETS standards as a model and to maintain its control over the ETS. Such a smaller linking partner fitted the EU's interests well (Interview 15). Thus, Switzerland figured as the ideal partner because its influence in the linked market was expected to be very small.

For Swiss policy-makers, this power imbalance, and that the SW ETS could join the EU ETS if they accepted the overall conditions and regulations, had been clear from the beginning of the linking considerations (Interview 16). The size of the EU ETS was not perceived as challenging by Swiss policy-makers; to the contrary, it was what Switzerland was looking for (Interview 10). Switzerland accepted the conditions and coped with the European model, with the exception of EU aviation regulations, where Switzerland raised major concerns (Interviews 2, 16). This shows that the acceptance of the power imbalance was not as undisputed as it may have appeared. Switzerland and the EU were divided over the treatment of aviation emissions under the linked carbon market. The conflict originated from the EU's attempt to treat flights between the EU and Switzerland as intra-European flights and thus regulate them under the EU ETS, while other EU-external flights were temporarily exempted from the system in order to give the International Civil Aircraft Organization (ICAO) time to design a global mechanism.[5] This was seen as a political affront by Switzerland and it demanded to be treated as a non-European country until the linking deal between the jurisdictions was concluded (Aviation Week, 2013a, Aviation Week, 2013b). The EU was forced to make this concession and apply the same rules to Swiss aviation companies as to other non-European countries. For Europe, it was very important to include the Swiss aviation sector into the common market (Interviews 10, 13, 16). This would allow the EU to close a loophole that allowed central European airports, such as Zurich and Basel, to circumvent EU carbon market regulations (Interview 16). This situation incentivized the EU to move forward with the link to Switzerland (Interview 16), because it had not only represented a regulatory gap, that made the regulation of the aviation sector in the EU less efficient, it also could have been perceived as a lack of political strength and influence. For Switzerland, the inclusion of aviation to a common market was challenging, because this sector had not faced emissions regulations related costs before. Thus, from an economic perspective the inclusion of the aviation sector was expected to bring additional costs to the Swiss aviation industry (Interviews 13, 16). Overall, the inclusion of aviation is perceived to have significantly prolonged the negotiations (Interview 16, 13, 10). Ultimately, however, it did not stop them. Swiss policymakers eventually agreed to adopt the EU ETS standards, amend its legislation, and include aviation into the common market.

Even though this linking process can be claimed as successful, some elements, mostly *structural aspects related to the legislative process, decision making*, and *consent structure in the partner jurisdictions*, have had inhibiting impacts. The enactment of an international treaty follows a very formal procedure and for that reason the negotiations between both jurisdictions had very formal character (Interview 13, 16). In the EU, the complex, time-consuming decision-making process slows down negotiation processes (Interviews 15, 16). This can be explained by the need to constantly involve the 27 member states, inform them, and hold regular discussion meetings (Interview 16). This process is prone to conflicts. For example, during the linking negotiations, member states started to feel discomfort with their involvement, perceiving uncertainties with specific topics, such as the provisions on delinking. Consequently, they demanded a better information flow from the EU COM (Interview 16). This situation was even intensified by the fact that the EU often did not prioritize linking to Switzerland on the political agenda. The reform of the EU ETS has clearly been the number one priority during the negotiation phase with Switzerland and EU officials argued that it consumed significant attention from the personnel and resources, leaving little capacity for other topics (Interviews 11, 15, 16). In contrast, a prioritization by the Swiss administration helped to push negotiations forward (Interview 13).

However, another, more prevailing hurdle appeared in the above-mentioned Swiss referendum on migration (Interviews 10, 11, 13, 15, 16). Effectively, the EU put linking negotiations on ice in reaction to the public vote, even though most technical aspects had been successfully negotiated (Interview 15). The stop of linking negotiations can be considered a negative by-product of these more general political discussions (Interview 13, 16). By deciding over the need to regulate EU immigration to Switzerland in a public referendum, the Swiss public to some degree questioned its ties to the EU and asserted greater independence. The referendum clearly slowed down and temporarily stopped the linking process (Interviews 10, 13, 15, 16), casting doubt on the potential success of that link at the time (Interview 6). Policy-makers on both sides could not ignore this domestic development; this conflict was only resolved through a legislative reaction by Switzerland in the form of a proposal regulating migration.

Even though the EU ETS implementation process had brought a large array of constituents to the table, in this linking process EU *stakeholder activity* was limited, with interest groups neither advocating for nor against a link. On the Swiss side, the situation was more complex. A strong push by the responsible agency, BAFU, and its then director was described as having advanced the link (Interview 13). The then-director of the BAFU announced his intention to leave toward the end of 2015 but showed a personal interest in finalizing the agreement while he still served in the agency (Interview 13). He invested significant personal efforts in the negotiations, for example through personal engagement with EU COM and by exercising pressure on the technical and working level. His efforts to find solutions to outstanding points helped finalize the agreement (Interview 13).

Furthermore, large parts of Swiss stakeholders from the industrial and service sectors supported linking to the EU ETS, for example, the federation of Swiss business 'economiesuisse' and the Swiss Alliance for Mountain Regions (Schweizerische Arbeitsgemeinschaft für die Berggebiete SAB, 2016, economiesuisse, 2016), because it represented a more economical alternative to the carbon tax (Interview 13). Yet, over the years of negotiation, stakeholder opposition also developed in Switzerland. For example, the Swiss aviation sector was more skeptical about the link and raised considerable resistance to its inclusion in the ETS (Interview 13).

In the important Swiss stakeholder consultation process ('Vernehmlassung') of September 2016, stakeholders helped to finalize a successful legislative text by providing comments and information (Interview 13). For example, environmental groups expressed the concern that the oversupply of emissions allowances in the EU ETS could risk the environmental integrity of the linked system (öbu, 2016, Klima Allianz Schweiz, 2016). The Schweizer Volkspartei (SVP), which is traditionally one of the strongest parties in Switzerland, displayed a general opposition against further association with the EU and directly opposed the proposed EU link, while also demanding a general abolition of the SW ETS (SVP, 2016).

In 2018, the pending consent of the Swiss Parliament on ratifying the agreement and amending the domestic legislation represented another opportunity for stakeholders to oppose the link. For example, the conflict between industry and the aviation industry reappeared. The Swiss liberal party ('Freie Demokratische Partei' (FDP)) requested to postpone the political discussion. They argued that the EU ETS and the common market's relationship with and participation in the planned ICAO mechanism should be clarified first;[6] this move signaled that they would make the whole linking agreement fail if the topic was not satisfactorily agreed upon (Forster, 2018). But also other topics, such as the then-higher EU allowance price entered the domestic discussion. While industrial groups such as economiesuisse and companies such as Holcim advocated for the link despite the prospective of higher costs, some political parties again showed strong opposition. The SVP, for example, criticized the imbalance of influencing power that favored the EU ETS (Häne, 2018).

Additional opposition came from the Green Party and parts of the social democratic party Sozialdemokratische Partei der Schweiz (SP); they argued that the ETS carbon price was too low and that the SW ETS should be abolished in general (Häne, 2018). The WWF Switzerland saw linking to the EU ETS as negative for climate policy, expressing the concern that linking would undermine the Swiss domestic climate policy target (WWF Schweiz, 2018). Behind this general opposition, however, stood their critique of the then low EU carbon price of approximately 10 EUR (Klima Allianz Schweiz, 2017, WWF Schweiz, 2017). This book's analysis found no information on a position change after the recent EUA price increase.

Overall, *international developments* seem to have played as very small role in the linking between the EU ETS and the SW ETS. The first informal linking talks between EU and Swiss policy-makers occurred at a time when a general optimism for linking existed worldwide, and also the disappointment after the failure of the negotiations at COP 15 in 2009 in Copenhagen did not stop linking efforts. A further prolongation of the negotiations was possibly caused by the EU's pursuit of a link to the AUS CPM during 2012–2014, a project that likely drew EU policy-makers' attention and resources away from linking negotiations with Switzerland (Interview 11).

Notes

1 96 CHF.
2 All foreign currencies (CHF) were converted by the author, as of 03.05.2020 (https://bankenverband.de/service/waehrungsrechner/).

3   125 CHF.
4   12.65 CHF.
5   The inclusion of flights into the EU ETS had led to several conflicts internationally. In a reaction to the worldwide opposition, the EU accepted to exclude 'non-intra-EU flights', from the EU ETS. This was a 'stop-the clock-procedure'.
6   The EU and Switzerland are signatories to the Carbon Offsetting and Reduction Scheme for International Aviation (CORSIA) under ICAO. Swiss constituents argued that with the EU ETS in place, there could be a double regulation of the aviation sector.

References

AVIATION WEEK. 2013a. *EU ETS: More controversy ... this time over Switzerland* [Online]. Available: http://aviationweek.com/blog/eu-ets-more-controversy-time-over-switzerland [Accessed 21.12.2016].

AVIATION WEEK. 2013b. *Switzerland opposes scope of ETS stop-the-clock proposal* [Online]. Available: http://aviationweek.com/awin/switzerland-opposes-scope-ets-stop-clock-proposal [Accessed 21.12.2016].

BAFU. 2009. *Bundesrat erteilt Verhandlungsmandat für Verknüpfung mit EU-Emissionshandel* [Online]. Available: www.bafu.admin.ch/klima/03449/12696/index.html?lang=de&msg-id=30717 [Accessed 20.09.2016].

BAFU. 2011. *Erste formelle Verhandlungen mit EU zur Verknüpfung der Emissionshandelssysteme* [Online]. Available: www.admin.ch/gov/de/start/dokumentation/medienmitteilungen.msg-id-38021.html [Accessed 20.09.2016].

BAFU. 2013. *Fünfte Verhandlungsrunde Schweiz – EU zur Verknüpfung der Emissionshandelssysteme* [Online]. Available: www.admin.ch/gov/de/start/dokumentation/medienmitteilungen.msg-id-51350.html [Accessed 20.09.2016].

BAFU. 2016. *Vernehmlassung vom 31.08.2016-30.11.2016 über die zukünftige Klimapolitik der Schweiz* [Online]. Available: https://biblio.parlament.ch/e-docs/387365.pdf [Accessed 08.04.2020].

BAFU. 2018. *Emissions trading* [Online]. Available: www.bafu.admin.ch/bafu/en/home/topics/climate/info-specialists/climate-policy/emissions-trading.html [Accessed 01.12.2018].

BAFU. 2019a. *Emissionen von Treibhausgasen nach revidiertem CO_2-Gesetz und Kyoto-Protokoll, 2. Verpflichtungsperiode (2013–2020)* [Online]. Available: www.bafu.admin.ch/dam/bafu/en/dokumente/klima/fachinfo-daten/CO2_Statistik.pdf.download.pdf/CO2_Publikation_de_2019-07.pdf [Accessed 02.04.2020].

BAFU. 2019b. *Emissionshandelssystem EHS für Betreiber von Anlagen* [Online]. Available: www.bafu.admin.ch/bafu/de/home/themen/klima/fachinformationen/klimapolitik/emissionshandel/schweizer-emissionshandelssystem--ehs-.html [Accessed 02.04.2020].

BAFU. 2019c. *Ratification of the agreement* [Online]. Federal Office for the Environment FOEN. Available: www.bafu.admin.ch/bafu/en/home/topics/climate/info-specialists/climate-policy/emissions-trading/linking-the-swiss-and-eu-emissions-trading-schemes/ratifikationdesabkommens.html [Accessed 07.01.2020].

BAFU. 2019d. *CO2 levy* [Online]. Available: www.bafu.admin.ch/bafu/en/home/topics/climate/info-specialists/climate-policy/co2-levy.html [Accessed 02.01.2020].

BUNDESVERSAMMLUNG SCHWEIZ. 2020. *Ausgewogenes CO$_2$-Gesetz und Solar-Offensive* [Online]. Bern. Available: www.parlament.ch/press-releases/Pages/mm-urek-n-2020-02-12.aspx?lang=1031 [Accessed 08.04.2020].

CARBON MARKET WATCH. 2015a. *Towards a global carbon market: Prospects for linking the EU ETS to other carbon markets* [Online]. Available: https://carbonmarketwatch.org/wp-content/uploads/2015/05/NC-Towards-a-global-carbon-market-report_web.pdf [Accessed 02.04.2020].

COUNCIL OF THE EUROPEAN UNION. 2019. *Linking of Switzerland to the EU emissions trading system – entry into force on 1 January 2020* [Online]. Available: www.consilium.europa.eu/en/press/press-releases/2019/12/09/linking-of-switzerland-to-the-eu-emissions-trading-system-entry-into-force-on-1-january-2020/ [Accessed 02.04.2020].

ECONOMIESUISSE. 2016. *Klimapolitik der Schweiz nach 2020: Stellungnahme zur Vernehmlassung* [Online]. Available: www.economiesuisse.ch/sites/default/files/publications/Stellungnahme%20inklusive%20Anhang.pdf [Accessed 08.04.2020].

ECOPLAN. 2008. *Volkswirtschaftliche Auswirkungen von CO$_2$-Abgaben und Emissions-handel für das Jahr 2020. Analyse der volkswirtschaftlichen Auswirkungen mit Hilfe eines allgemeinen Mehrländer-Gleichgewichtsmodell* [Online]. Bern: Study commissioned by the Federal Office for the Environment. Available: www.ecoplan.ch/download/co2g_sb_de.pdf [Accessed 03.04.2020].

EDF & CDC CLIMAT RESEARCH. 2015. *Switzerland: An emissions trading case study* [Online]. Available: www.ieta.org/resources/Resources/Case_Studies_Worlds_Carbon_Markets/switzerland_case_study_may2015.pdf [Accessed 01.03.2019].

EU COM. 2010. *Opening negotiations with Switzerland on linking emissions trading systems* [Online]. Available: http://ec.europa.eu/clima/news/articles/news_2010110501_en.htm [Accessed 19.09.2016].

EU COM. 2019a. *Bericht der Kommission an das Europäische Parlament und Den Rat Bericht über das Funktionieren des CO$_2$-Marktes der EU* [Online]. European Commission. Available: https://eur-lex.europa.eu/legal-content/DE/TXT/PDF/?uri=CELEX:52019DC0557&from=EN [Accessed 02.04.2020].

EU COM. 2019b. *Agreement on linking the emissions trading systems of the EU and Switzerland* [Online]. European Commission. Available: https://ec.europa.eu/commission/presscorner/detail/en/IP_19_6708 [Accessed 02.04.2020].

EU COM. 2019c. *FAQs on Swiss linking* [Online]. Available: https://ec.europa.
eu/clima/sites/clima/files/ets/markets/docs/faq_linking_agreement_part2_
en.pdf [Accessed 07.01.2020].

EU PA. 2014. *Briefing: The state of EU-Switzerland relations in the EMPL areas
of responsibility* [Online]. Available: www.europarl.europa.eu/RegData/
etudes/BRIE/2014/536313/IPOL_BRI(2014)536313_EN.pdf   [Accessed
21.09.2016].

EUROPEAN VOICE. 2014. *EU adopts mandate for Switzerland talks* [Online].
Available: www.politico.eu/article/eu-adopts-mandate-for-switzerland-talks/
[Accessed 19.09.2016].

FORSTER, C. 2018. *FDP Verzoegert Verknüpfung der Emissionshandelssysteme*
Bern: Neuer Zürcher Zeitung. [Online]. Available: www.nzz.ch/wirtschaft/
fdp-verzoegert-verknuepfung-der-emissionshandelssysteme-ld.1377602
[Accessed 09.03.2019].

GLOCK, D., WALLNER, K., RUNGE, P., ERLING, U. M. & RIEDINGER,
L. 2016. *Implikationen der Marktregulierung im Kohlenstoffmarkt*
[Online]. Dessau: Umweltbundesamt (UBA) CLIMATE CHANGE 04/
2016. Available: www.umweltbundesamt.de/sites/default/files/medien/
378/publikationen/climate_change_04_2015_implikationen_der_
marktregulierung_im_kohlenstoffmarkt.pdf [Accessed 03.04.2020].

GROß, T. 2011. Klimaschutzgesetze im Europäischen Vergleich. *Z U R.*

HÄNE, S. 2018. *Dreckschleudern Sind Hier Billiger als in der EU – Schweiz*
[Online]. TagesAnzeiger. Available: www.tagesanzeiger.ch/schweiz/standard/
warten-bis-zur30eurogrenze/story/23059789 [Accessed 12.12.2019].

ICAP. 2020b. *Swiss ETS. ETS detailed information* [Online]. Available: https://
icapcarbonaction.com/en/?option=com_etsmap&task=export&format=pdf
&layout=list&systems[]=64 [Accessed 08.04.2020].

ICAP. 2020c. *EU Emissions Trading System (EU ETS). ETS detailed information* [Online]. Available: https://icapcarbonaction.com/en/?option=com_
etsmap&task=export&format=pdf&layout=list&systems[]=43   [Accessed
08.04.2020].

KLIMA ALLIANZ SCHWEIZ. 2016. *Klimapolitik der Schweiz nach 2020:
Übereinkommen von Paris, Abkommen mit der Europäischen Union über
die Verknüpfung der beiden Emissionshandelssysteme, Totalrevision des
CO_2-Gesetzes* [Online]. Available: www.sp-ps.ch%2Fsites%2Fdefault%2F
files%2Fdocuments%2F16-291_co-2_gesetz_-_klimapolitik_nach_paris._
fragekatalog.pdf [Accessed 08.04.2020].

KLIMA ALLIANZ SCHWEIZ. 2017. *Argumentarium gegen die Verknüpfung
der Emissionshandelssysteme der EU und der Schweiz. Klimaschutzregulierung
nur wenn es etwas nützt* [Online]. Available: www.wwf.ch/sites/default/files/
doc-2017-11/171123_FaktenblattLinkingEU_CH_ETS.pdf   [Accessed
23.4.2019].

NEUE ZÜRCHER ZEITUNG. 2009. *Der Emissionshandel – ein umstrittenes
Instrument des Klimaschutzes* [Online]. Available: www.nzz.ch/der_
emissionshandel__ein_umstrittenes_instrument_des_klimaschutzes-
1.2047485 [Accessed 03.10.2019].

OBERAUNER, I. M. & KRYSIAK, F. C. 2008. *Nutzen und Kosten einer klimapolitischen Kooperation zwischen der Schweiz und der EU* [Online]. Basel: University of Basel, Center of Business and Economics (WWZ), WWZ Forschungsbericht, No. 02/08. Available: www.econstor.eu/bitstream/10419/127520/1/wwz-fb-2008-02.pdf [Accessed 12.03.2018].

ÖBU. 2016. *Klimapolitik der Schweiz nach 2020: Übereinkommen von Paris, Abkommen mit der Europäischen Union über die Verknüpfung der beiden Emissionshandelssysteme, Totalrevision des CO_2-Gesetzes* [Online]. Available: www.oebu.ch/admin/data/files/section_asset/file_de/1443/vernehmlassungsantwort-klimapolitik-nach-2020-oebu-final[1].pdf?lm=1480947026 [Accessed 08.04.2020].

RUTHERFORD, A. P. 2014. Linking emissions trading schemes: Lessons from the EU-Swiss ETSs. *Carbon & Climate Law Review*, 8, 282–290.

SANTIKARN, M., LI, L., THEUER, S. L. H. & HAUG, C. 2018. *A guide to linking emissions trading systems* [Online]. Berlin: Report published by ICAP. Available: www.icapcarbonaction.org/publications [Accessed 03.09.2019].

SCHWEIZERISCHE ARBEITSGEMEINSCHAFT FÜR DIE BERGGEBIETE SAB. 2016. *Fragen an die Vernehmlassungsteilnehmenden Klimapolitik der Schweiz nach 2020* [Online]. Available: www.sab.ch/fileadmin/user_upload/customers/sab/Stellungnahmen/2016/CO2-Gesetz_Fragekatalog_SAB.pdf [Accessed 08.04.2020].

SVP. 2016. *Klimapolitik der Schweiz nach 2020* [Online]. Available: www.svp.ch/partei/positionen/vernehmlassungen/klimapolitik-der-schweiz-nach-2020/ [Accessed 08.04.2020].

THE FEDERAL COUNCIL GOVERNMENT OF SWITZERLAND. 2011. *Federal Act on the reduction of CO_2 emissions (CO_2 Act)* [Online]. Available: www.admin.ch/opc/en/classified-compilation/20091310/index.html [Accessed 13.09.2019].

WWF SCHWEIZ. 2017. *EU-Emissionshandelssystem bringt Buerokratie statt Klimaschutz* [Online]. Available: www.wwf.ch/de/medien/eu-emissionshandelssystem-bringt-buerokratie-statt-klimaschutz [Accessed 03.04.2020].

WWF SCHWEIZ. 2018. *Emissionshandel mit der EU: Rosinen picken, aber richtig* [Online]. Available: www.wwf.ch/de/medien/emissionshandel-mit-der-eu-rosinen-picken-aber-richtig [Accessed 30.04.2020].

6 The EU Emissions Trading System and the Australia Carbon Pricing Mechanism

An agreed-upon, but unrealized link

Climate policy in Australia and the implementation of the AUS CPM

Research predicts that climate change will hit Australia stronger than other parts of the world, with up to a 5.1°C temperature rise estimated by 2090 (CSIRO & the Bureau of Meteorology, n.d.). Further developments in Australia, such as the deterioration of the largest living structure, the Great Barrier Reef, make climate change increasingly tangible (Australian Government, n.d.-b). Even though in terms of absolute emissions Australia is not among the largest emitters worldwide, with 533.6 Mt CO_2e emitted in 2015 (excl. LULUCF), its per capita emissions of 22.1 t CO_2e per person (2015) are very high compared to global averages (Commonwealth of Australia, 2017). In Australia, the energy sector is responsible for the largest share of emissions, approximately 78.7% (Commonwealth of Australia, 2017), because of its reliance on coal for domestic electricity generation and the strong coal, natural gas, uranium and minerals export sectors (Bailey et al., 2012). Under the UNFCCC and in the NDC prepared in August 2015, Australia committed to an economy-wide target to reduce greenhouse gas emissions by 26–28% below 2005 levels by 2030 (Australian Government, 2015c). During the observed time of this case study, a target of reducing its greenhouse gas (GHG) emissions to 5% below 2000 levels was effective (Jotzo, 2012).

Despite Australia's strong exposure to climate change, climate policy has gone down a very curvy road in Australia. Not only has the climate policy mix been altered with almost every new government over the past two decades, also stakeholders' and the public's positions have oscillated back and forth between climate activism and skepticism. Carbon pricing is among the policies that have experienced strong rises and falls over the years. Reflecting briefly on the history of ETS implementation helps to understand why ETS had become such a politically sensitive topic.

The first discussions on carbon pricing, including ETS, are recorded to have taken place already in the early 1990s (Jotzo, 2012). However, they met strong opposition from several crucial businesses from the energy sector (Hudson, 2015). In 2000, a first ETS plan was proposed to the Senate (Parliament of Australia, 2000) while in parallel, Australian states, such as New South Wales, introduced carbon pricing mechanisms (Quiggin et al., 2014).[1] On the national level, in 2003, then-Prime Minister John Howard vetoed a further ETS proposal (Minchin, 2006). However, he changed his position in 2006, giving ETS another chance (ABC News, 2007). By 2007, public awareness of climate change had started to grow (Crowley, 2013) and it became a decisive topic in the elections, ultimately resulting in a change in government (Crowley, 2013). In 2007, the government of Kevin Rudd ratified the Kyoto Protocol and, in December 2008, it presented a detailed proposal for the 2010 introduction of an ETS: the 'carbon pollution reduction scheme'. However, the proposal became increasingly unpopular, with two failed attempts to get ETS legislation through Parliament. Rudd announced the withdrawal of the ETS in 2010 (Bailey et al., 2012) and then delayed carbon pricing plans until 2013 (Herald Sun, 2010). The proposed intro-duction of the carbon pricing mechanism had reached the center of a politically heated and even 'toxic' discussion on climate politics (Bailey et al., 2012). On one side, green and ecologic movements criticized the planned compensation and free permits provisions to large industrial emitters. On the other side, some industrial groups, such as the power sector and coal mining industries, the Business Council of Australia, the Minerals Council of Australia and the Australian Industry Group (Taylor, 2014, White, 2013), as well as opposition parties foresaw severe economic impacts and therefore rejected ETS (Beeson and McDonald, 2013). In the media, industrial lobby groups raised the fear that the AUS CPM would increase costs for consumers (such as electricity costs) and argued that the mechanism would cause firm closures and job losses (Lewis and Jacob, 2013).

In spite of these conflicts, in 2012, the Clean Energy Act 2011 (Australian Government, n.d.-a), came into force with a carbon pricing mechanism as its centerpiece: the AUS CPM (Jotzo, 2012). The admin-istration of Julia Gillard was forced to introduce a carbon price in order to be able to keep her minority government in office (Jotzo, 2012).[2] The politically contemptuous atmosphere around climate policy had not calmed down and the lack of bipartisan support for the AUS CPM generated an overall sense of uncertainty around the scheme (Jotzo, 2012). The opposition leader Tony Abbott made the repeal of the carbon price his number one electoral topic and started a public cam-paign against the mechanism (Taylor, 2014). Leading up to the election, public polls indicated that about 52% of the Australian public opposed the proposed scheme (Irigoyen, 2017). When Tony Abbott won the subsequent elections in 2013, he immediately introduced draft legislation, the Direct Action Plan, to repeal the AUS CPM. In 2014 this legislation was adopted, resulting in the repeal of the AUS CPM.

Having gone into effect on July 1, 2012, the AUS CPM had been designed to be the centerpiece of a policy mix which contained measures for energy efficiency, renewable energies, energy innovation and a carbon farming initiative (Banerjee, 2012). The ETS aimed at contributing to Australia's GHG reduction target of a 5% reduction from 2000 levels by 2020 (Jotzo, 2012).[3] The AUS CPM started operation with a three-year fixed price period (2012–2015). During the first 'carbon tax' phase, unlimited permits were allocated at a fixed price of approximately 13.5 EUR in the first year, thereafter increasing in price at a rate of 2.5% each year, reaching approximately 15 EUR in 2015 (Jotzo, 2012, Banerjee, 2012).[4] In the second phase, starting in July 2015, the AUS CPM would have transitioned to a full ETS with flexible prices. Prices were to be regulated through a price ceiling that would be set at 11.7 EUR and rise by 5% yearly.[5] The price ceiling was scheduled to expire by 2018 (Spencer et al., 2012).

The system covered direct carbon dioxide emissions from electricity generation, stationary energy, landfills, wastewater, industrial processes and fugitive emissions (Australian Government, 2015b). Roughly half of the covered emissions would have come from electricity generation from power stations using non-renewable fuels (Banerjee, 2012). Approximately 350 large emitters had obligations under the ETS. A liability threshold was set at the amount of 25,000t; entities with emissions below this number were not covered (Australian Government, 2015b). Overall, the AUS CPM covered approximately 60% of Australia's GHG emissions (Jotzo, 2012).

The AUS CPM handed out allowances for free to industries, such as the steel and aluminum sectors, and to companies that qualified either under the Jobs and Competitiveness Program or the coal-fired generation assistance (Jotzo, 2012).[6] These payments were large, amounting to 1.75 billion EUR[7] per year during the first fixed price phase (Jotzo, 2012).[8] The system also contained penalties, so-called unit shortfall charges, for those entities that did not surrender sufficient allowances for compliance. For the 2012–2013 financial year, the unit shortfall charge was of 14.5 EUR per unit (Australian Government, 2015c).[9] With the start of the second ETS phase, the government would have started auctioning off a certain amount of allowances (Jotzo, 2012)

From the flexible phase on, entities would have been able to sur-render up to 50% in international units (Australian Government, n.d.-a). Entities could also use domestically-sourced offsets from the Carbon Farming Initiative without restriction (Peel, 2014). The credits currency under the AUS CPM was named Australian Carbon Credit Units (ACCUs). Another important feature of the AUS CPM was its approach to recycle the financial revenues: More than half of the projected revenue (2.9–5.26 billion EUR) were to be returned as lower income taxes or higher welfare payments, giving advantages to lower-income households (Jotzo, 2012).[10] The Australian Government coupled the mechanism with several support and compensation programs, such as pension increases for families and students; a program for emissions-intensive, trade-exposed industries; a special program for the steel industry; and the Energy Security Fund. They also established an inde-pendent Climate Change Authority (AUS CCA) (AustralianPolitics. com, 2011). The AUS CCA figured as the main point of contact and responsible authority for ETS and linking.

Linking the EU ETS and the AUS CPM

It is difficult to identify the exact moment when linking considerations between Australia and the EU began. In Australia, ETS linking was recommended in the 2008 Garnaut Report and the government's white paper (Garnaut, 2008) and bilateral links were discussed at least in early 2009 (Jotzo and Betz, 2009). Thus, unofficial talks to the EU on linking already started in parallel to the implementation process of the AUS CPM. In 2011, an intense exchange commenced and officials from the EU COM and the AUS CCA engaged in explorative talks on technical aspects and potential linking options (Santikarn et al., 2018).

In the beginning of 2012, Australian and EU policy-makers reached a shared understanding on linking the EU ETS and the AUS CPM (EU COM, 2012). They had negotiated an interim unilateral link to start in July 2015, which would turn into a full bilateral link by July 2018 at the latest (EU COM, 2012). During the first phase of the link, only Australian companies could use EUAs, meeting up to 50% of their liabilities through purchasing international units (EU COM, 2012). Both jurisdictions entered into formal negotiations for an international linking treaty in 2013 (EU COM, 2013, Talberg and Swoboda, 2013). Nevertheless, the negotiations were put on ice the same year, after the Australian general elections in September 2013 resulted in a government turnover (Interview 16). The then-incoming administration repealed the AUS CPM in 2014 and linking talks were abandoned (EU COM, n.d. a).

Several design differences between the EU ETS and the AUS CPM can be identified. Initially, the most notable difference was the planned price containment measures of the AUS CPM, namely the price floor and ceiling (Carbon Market Watch, 2015a). The price floor would have been undermined by the much cheaper EUAs, rendering it ineffective. This incompatibility was fixed when the Australian government agreed to remove the price floor for the ETS phase of the AUS CPM in 2015 and the price ceiling in 2018. Additionally, the Australian domestic offset scheme represented a challenge to the EU. The Carbon Farming Initiative was based on the generation of credits from agricultural and land-use management, sectors excluded under the EU ETS. The resolution of this issue was postponed in discussions, remaining open until the termination of linking negotiations (Carbon Market Watch, 2015a) (Interview 18). The EU ETS and the AUS CPM also differed slightly in their coverage, since the AUS CPM covered methane emissions, but the EU ETS did not. Overall, many technically unclear aspects were left for further negotiations in order to get the agreement finished quickly. However, they would have likely come up as challenges if negotiations had gone further (Interviews 16, 18).

Linking activities between the EU and Australia in the time period of approximately 2011–2014 can be roughly situated in the beginning of what this book conceptualizes as the second phase of linking; in this stage, negotiations start to be more formal and officially envision the goal of realizing a treaty.

The data evaluated for this research shows that this linking initiative was strongly motivated by *economic interests*. On the Australian side, it was a priority to lower the overall costs of carbon pricing by linking to the EU ETS and integrating international credits into the AUS CPM. This acted as one of the main driving factors for Australian policy-makers to initialize linking (Interview 9, 16). During the examined time period, EUA prices already showed a downwards trend and remained relatively low. Thus, the EU ETS was seen as a promising means to access to more and cheaper abatement opportunities in comparison to the fixed Australian price. At that time, the expected allowance price for the AUS CPM's future market period was 29 EUR; this suggested an increasing need for industry to access a larger and cheaper market in order to lower companies' compliance costs (Interview 9).[11] Additionally, AUS CCA realized a study, which found that a mix of domestic and international emissions reductions would be much cheaper than the use of domestic allowances only. It estimated a price of 38 EUR/t in 2020 if Australia were to achieve the 5% target with only domestic efforts (AUS CCA, n.d.). In comparison, the study estimated a price of only 17.6 EUR/t for 2020 if it were to integrate credits from the EU ETS and/or the CDM (AUS CCA, 2014).[12] Policy-makers were considering linking, because they saw Australia as a country with high abatement costs due to the high emissions factor in their energy mix (Interview 18). Addressing competitive concerns through the levelling of carbon market policies was also a main motivation (Interview 2) (AUS CCA, n.d.). For EU policy-makers, economic interests seem to have played only a minor role in this linking activity. The justification that EU policy-makers were motivated by an interest in reducing the oversupply of EUAs in the EU ETS by introducing demand from Australian companies is lacking, as this would not have brought concrete benefit for EU ETS covered companies. This new demand was seen as more of an additional benefit and not the main motivation for linking (Interview 16).

In general, the EU and Australia saw ETS as a priority instrument for achieving *climate targets* and both framed their ETSs as their main climate policy tool (Interview 9). Nonetheless, during the negotiations, concerns about the EU ETS's *environmental integrity* existed on the Australian side (Interviews 11, 18). The developments of a price collapse, allowance oversupply and the then-consistently low prices in the EU ETS raised concerns in Australia (Interviews 11, 18). In Australia, a general discussion can be observed on the advantages and disadvantages of international credits within the AUS CPM. Some stakeholders were wary of the inclusion of international credits, seeing it as a distraction from the real challenge in Australia: the domestic reduction of emissions (Interview 18). Indeed, a cheap carbon price post-linking would probably not have been able to change the coal-dependent energy structure in Australia, and therefore would not have helped with Australia's domestic environmental challenge of achieving a less carbon-intensive economy. On the other side, EU policy-makers held a very positive view of Australian ambition; this opinion, together with the perceived high quality of the Australian approach, was a driver for linking (Interview 15, 16). Nonetheless, both climate policy approaches were perceived as very different (Interview 18) not only with respect to their emitting sectors (the AUS CPM included methane and the EU ETS did not), but also to their political context, such as their cost abatement curves. The caveats of these differences led to uncertainty on both sides (Interviews 18, 11).

Even though the interviews realized for this work have not highlighted how the expected *loss or gain of co-benefits* affected this linking attempt, one aspect can be noted. Linking to the EU ETS would have implied a revenue loss for the Australian Government, because the EUAs prices were lower than the carbon price under the AUS CPM. This financial loss could have become an important political burden, since a large part of these revenues in Australia was attached to adjacent programs or redistributed to energy consumers. Furthermore, the program's revenue was seen as a way to increase support for and ultimately lock in the AUS CPM in the climate policy mix (Interview 9). It is difficult to estimate whether the potential loss of financial revenues acted as a hurdle during the negotiations. Policy-makers concentrated on highlighting the gain of co-benefits such as the expected decrease in electricity costs such as electricity prices, which was seen as an opportunity to console stakeholder concerns (Interview 18).

Also in this case, the analysis has shown that the demonstration of *leadership* and the feasibility of linking played a central role. In the EU, the general possibility of an intercontinental link with another large jurisdiction that depended strongly on fossil fuels was perceived as internationally meaningful and, as one interviewee stated, it would have raised attention simply for being the first of its kind (Interview 15). This link would have given a very positive signal to the world in favor of a global carbon market, initiated by the EU (Interview 16). Consequently, it would strengthen the international perception of the EU as a climate frontrunner. Linking could have underscored the EU's capacity to set standards, in spite of the EU ETS's oversupply and environmental integrity challenges. Such an agreement could have been the starting point for the development of an ETS club or the possibility for other jurisdictions to join the common market (Interview 15). Australian policy-makers were looking for a linking partner because they wanted to increase the value of their carbon pricing instrument and generate more support from Australian stakeholders (Interview 18). Through linking, the Australian administration could have proven their capability for realizing carbon pricing (Interview 18).

This analysis indicates that an *imbalanced power constellation* and the *difference in ETS size* have had both a pushing and an impeding effect on the linking initiative. Altogether, there was a stronger interest in linking coming from Australian policy-makers compared to EU policy-makers, since linking was more important for Australia than for the EU (Interview 2), for three reasons: First, Australian policy-makers expected increasing climate abatements costs for Australian companies. Second, the access to the cheaper EU ETS was an argument that policy-makers needed to hold on to in order to comfort industrial and public concerns and, ultimately, to stabilize the AUS CPM in the policy mix. Third, Australian policy-makers saw this international link as an opportunity to save the AUS CPM from its looming repeal after the general elections, as discussed in the following sections. These aspects made linking a much more substantial issue for the AUS CPM than the EU ETS, and probably put Australian policy-makers in a weaker negotiation position, leaving them more willing to accept EU conditions. This dynamic was visible, for example, in how quickly Australian policy-makers accepted the deactivation of the AUS CPM's price floor and a limitation on the use of offset credits, in order to progress with linking as quickly as possible. Also, the larger size of the EU created an imbalance of power between both jurisdictions. In the EU, being the larger ETS and being in a position of power were regarded as important elements in facilitating linking efforts (Interview 15). Yet, policy-makers' and stakeholders' concerns about losing regulatory control over the AUS CPM permeated the negotiations. For example, EU policy-makers perceived discussions in the Australian Parliament to have revolved around the fear of becoming the receiver of EU decisions (Interview 15). Even though Australian policy-makers were aware of this risk, these concerns, including those political priorities that came with the price floor, were sacrificed for the greater aim of linking and the associated cost benefits for companies (Interview 9).

The data collected for this book has shown that *systemic aspects*, such as the *structure of the legislative* and *political processes* as well as the setting of the *political agenda* played a determinant role in this case, exercising both positive and negative influences on linking. Underpinning the negotiations was the political uncertainty around the AUS CPM's future.

On the one hand, the anticipation of the general elections in 2013 in Australia provided a positive push for linking. In the same year as the launch of the AUS CPM, experts were already anticipating the possibility of its repeal as a result of the potential government turnover (Jotzo, 2012). Consequently, the scheduling of general elections gave policy-makers an extra incentive to finish linking negotiations (Interviews 11, 15, 16). They expected that the risk of breaking an international agreement would make the next government hesitate to withdraw policies internally (Interviews 2, 11, 18) (Müller and Slominski, 2016). This was an explicit strategy taken by the negotiators and linking became a *top priority on the political agenda* for Australian negotiators (Interview 18). Essentially, linking before the elections was seen as a safeguard for the AUS CPM's existence (Interview 6) and thus became rather a matter of 'politics' than 'policy' (Interview 18). Experts estimated that this situation explains why Australian negotiators were inclined to accept even those EU conditions, such as the abandonment of the price floor, which from an academic and practical perspective were regarded as problematic for linking (Interviews 6, 16).

On the other hand, the outcome of the elections, which resulted in a government turnover, did indeed lead to the ETS's repeal and an ultimate failure of the linking efforts (Interview 16). Yet, it was not only the politics of the new government and stakeholder activity that made linking fail, but the rather unfortunate timing of the elections. In other words, Australian negotiators ran out of time (Interview 18). The argument was raised, that with a different timing and had a carbon price been in existence a few more years, Australian stakeholders might have started to support more carbon pricing and climate policy. Possibly then, the ETS would have been kept and the jurisdictions would have proceeded with linking (Interview 18). This argument may be supported by public opinion polls that indicated a rise in general support for climate policy after 2014 (Irigoyen, 2017).

On the European side, *political systemic* or *legislative aspects* seem to have played a minor role. The EU COM initiated the linking talks and then involved member states officially, and through informal talks and meetings (Interview 16). However, negotiations advanced very quickly; they were also held to some degree behind closed doors and, for some time, at a high political level (Interview 2). Some member states felt that they were insufficiently informed or not involved in a timely manner (Interview 16). This culminated in an institutional dispute between EU COM and EU COU, in which the EU COM filed a legal suit against the EU Council, complaining about overly strict reporting rules (Interview 16) (EU Lex, 2015, Sabin Center for Climate Change Law and Columbia University, 2015). Even though interviewees have not named this situation as a concrete barrier to the linking activity, it probably added to the overall uncertainty around the future of this linking activity.

From a legal perspective, Australia and the EU had an easier start in the linking negotiations, because both were using emissions reductions currencies accepted under the Kyoto Protocol. This facilitated linking significantly, establishing a common ground on the technical side (Interview 16). It would have enabled the allowance exchange to be integrated into future domestic legislation as well as each country's accounting under the Kyoto commitments.

As some of the above paragraphs have already implied, this linking process was significantly affected by *stakeholders' opposition and advocacy,* first and foremost in Australia. A driving factor was the existence of several large Australian industrial and business groups who had a very strong interest in the cost efficiency aspect of linking, as they expected linking to the EU ETS would lower their costs (Interview

9) (Santikarn et al., 2018). Lowering costs was especially important for them, because they feared competitive disadvantages arising from the fact that Australia's important trade partners, such as China, Japan, and the United States, did not have carbon prices (Interview 18). Additionally, some large Australian companies with operations or head-quarters in the EU were obliged to participate in the EU ETS. For them, the prospect of transferability of credits across schemes was enticing; it would have facilitated compliance (Interview 18).

Nonetheless, not all Australian stakeholders advocated linking to the EU ETS. Some business groups such as the Cement Industry Federation expressed a generally positive view on linking, but would have pre-ferred geographically closer trading partners in the South Pacific region and raised concerns over the expected dominance of the EU over a linked market (Cement Industry Federation, 2013). Other Australian stakeholders, led by the power sector and coal mining industries, opposed the ETS policy and/or linking to the EU ETS. Groups, such as the Business Council of Australia, the Minerals Council of Australia and the Australian Industry Group opposed the AUS CPM publicly and financially supported Tony Abbott's Liberal's Party coalition against the AUS CPM (Taylor, 2014, White, 2013). These stakeholders had raised the argument that the AUS CPM would increase electricity, fuel and grocery prices and interest rates (The Climate Institute, 2013); they also argued that jobs would be lost through the closure of firms as a result of the mechanism (Lewis and Jacob, 2013). However, many of these attacks were targeted at the AUS CPM in general and less so to stop the linking endeavor in particular. Specifically for linking, the differences of the systems, such as the inclusion of methane in the AUS CPM, were perceived as challenging, because they raised distributional concerns, with Australian companies fearing a disadvantage (Carbon Market Watch, 2015a).

Additionally, the observed weaknesses of the EU ETS, such as the allocation oversupply, were highlighted by the political opposition and some stakeholders used this situation politically against linking (Interview 18). Some stakeholders, including academics and environ-mental groups such as the WWF, were concerned about the integrity and stringency of the EU ETS; this fear was rooted in the oversupply and windfall profits, which they perceived as an indication for a lack of ambition and as a risk to the effectiveness of the AUS CPM (Interview 18) (WWF Australia, 2012, Palmer, 2012, Rourke, 2012). They also expected that the potential high demand for cheaper EUAs would result in greater emissions reductions outside than inside Australia (WWF Australia, 2013). This view was reflected in discussions on the need for domestic emissions reductions in place of international credits (Interview 18). However, positions from environmental groups were not entirely negative; linking in general was welcomed (WWF Australia, 2012), in spite of the above noted concerns. The Green Party also welcomed the link to the EU ETS (Swoboda et al., 2012).

In Australia, carbon pricing in general was highly politicized and reached the top of the political agenda, fostering widespread awareness of the AUS CPM among the Australian public (Interview 2, 18, 9). At least four factors made carbon pricing unpopular within the Australian public: First, declining public support for climate policy as a whole (Irigoyen, 2017); second, the growing feeling of mistrust after the reversal of the government's initial promise that they would not implement a carbon tax, coupled by a general discomfort with 'another tax' and rising electricity prices (Interview 18) (Lewis and Jacob, 2013, Tennant, 2013); third, the huge media debate that reported disproportionately on the potential negative consequences of the AUS CPM (Tennant, 2013, Lewis and Jacob, 2013); and fourth, the opposition party's campaign against the carbon tax. Overall, high degree of uncertainty on the AUS CPM's future and lack of qualitative information on linking added to the overall negative attitude amongst the public (Interview 18).

In Australia, also political parties, but also some entrepreneurs and political leaders from the involved institutions, played a significant role for linking. The Australian Liberal Party made the abolition of the AUS CPM its number one topic and legislative priority in its electoral campaign (White, 2013). The political opposition leader and later prime minister of Australia, Tony Abbott, contributed to the emotionally-heated atmosphere by declaring a 'blood oath' to 'ax the tax' (Taylor, 2014). Conversely, the AUS CCA was perceived to have acted as a strong driving factor in favor of this link. Also the role of the Australian minister for climate change policy, Greg Combet, was highlighted as having been very positive and entrepreneurial in the linking dialogue (Interview 15). His commitment brought linking forward and his personal attachment and knowledge on ETS helped to move linking to a higher political level (Interview 15).

In the EU, industrial stakeholders were reported to have held an overall positive attitude toward the idea of a link. The German central industrial alliance BDI, for example, reports encouraging the EU COM on the linking activity, in part because they did not expect the AUS CPM to have a strong influence on the EU ETS (Interview 5). Nonetheless, this analysis did not find strong EU industrial stakeholder activity. Some voices were raised on the need to ensure environmental effectiveness. However, these were generally referring to the design flaws experienced by the EU ETS itself rather than the AUS CPM (WWF EU and Van den plas, 2012).

Contrary to the above-conceptualized factor of *proximity* between linking partners as a driver for linking, in this case, some interviewed experts have claimed that *geographical distance* created a positive environment for the negotiations between the EU and Australia (Interviews 15, 2). As large emitting industries in the EU, such as the cement and steel sectors, have few trade relationships with their Australian counterparts, they face less direct competition, which makes the topic less politically sensitive (Interview 15, 2). The lack of trading dependencies likely reduced concerns over distributional impacts through, for example, free allocation (Interview 15). Yet, the proximity between linking partners in terms of the quality of their relationship, which was described as very close and based on trust and mutual recognition, positively influenced linking progress (Interviews 15, 16). Especially trust, built through in-person meetings and talks, a close relationship and knowledge exchange, acted as an accelerator and path smoother for this link (Interview 16). Moreover, had talks continued, additional technical exchange might have taken place through fora such as ICAP and IETA, of which both jurisdictions are members, thereby helping further intensify their relationship.

Overall, Australian policy-makers' predisposition for an *international integration* of the AUS CPM facilitated linking efforts (Jotzo, 2012). Additionally, a beneficial starting point was guaranteed by the fact that both jurisdictions had ratified the Kyoto Protocol and needed to comply with their respective commitments (Interview 16).

The *broader developments of international climate policy* and the process under the UNFCCC show some more mixed impacts on this linking attempt. While a general disappointment with the process under the UNFCCC after the failure of the Copenhagen COP might have redirected attention to bilateral cooperation and linking, the fact that only very few countries were still committed to the Kyoto Protocol in its second period after 2012 probably caused additional pessimism. For example, experts observed that the Australian public feared that Australian 'unilateral' action would be economically disadvantageous and ineffective (Jotzo, 2012).

In parallel, the fast spread of carbon pricing policies worldwide also inspired Australia to initiate talks with the EU (Interviews 9, 18). Additionally, for EU policy-makers, the abandoning of talks with California and the certainty that no national US ETS would be launched made a link to the AUS CPM more appealing. Furthermore, in parallel to the linking attempt with the EU, in 2011, Australian policy-makers had considered a link to the New Zealand Emissions Trading Scheme, with a possible starting date in 2015 (ICTSD, 2011). In view of the close geographic and economic relationships between Australia and New Zealand, this could have been a logical step. However, Australian policy-makers' ultimate decision against linking the AUS CPM with the New Zealand Emissions Trading Scheme was taken as a symbol of trust by EU policy-makers, as it demonstrated a mutual commitment to linking (Interview 15).

Notes

1 New South Wales operated a carbon pricing system between 2003 and 2012, which was repealed when a national system started to operate.
2 The introduction of an initial fixed price phase was part of the coalition's compromise with Australia's Green Party (Jotzo, 2012).
3 Up to a 25% reduction, depending on other countries' international commitments.
4 Respectively, 23 AUD and 25.4 AUD.
5 20 AUD.
6 Coal-fired power stations would have received 3.21 billion EUR (5.5 billion AUD) over five years and coal mines 0.76 billion EUR (1.3 billion AUD) over six years (Jotzo 2012).
7 3 billion AUD.
8 All currencies (AUSD) were converted by the author, as of 03.05.2020 (https://bankenverband.de/service/waehrungsrechner/).
9 29.9 AUD.
10 . Respectively, 5–9 billion AUD.
11 50 AUD.
12 Respectively, 65 AUD and 30 AUD. These numbers can be found in a report that was released shortly before the AUS CPM was repealed. The medium scenario expects prices of 17.6 EUR (30.1 AUSD) in 2020, based on forecasts for the European credits' price and estimating an approximately 50% use of these credits (AUS CCA, 2014).

References

ABC NEWS. 2007. *Howard announces emissions trading system* [Online]. Available: www.abc.net.au/news/2007-07-17/howard-announces-emissions-trading-system/2505080 [Accessed 13.11.2017].

AUS CCA. 2014. *Reducing Australia's greenhouse gas emissions—Targets and progress* [Online]. Australian Government. Available: www. climatechangeauthority.gov.au/files/files/Target-Progress-Review/Targets%20and%20Progress%20Review%20Final%20Report.pdf [Accessed 02.05.2020].

AUS CCA. n.d. *Benefits and risks of using International Units* [Online]. Available: http://climatechangeauthority.gov.au/reviews/using-international-units-help-meet-australias-emissions-reduction-targets/benefits-and   [Accessed 03.04.2020].

AUSTRALIAN GOVERNMENT. 2015a. *Unit shortfall charges* [Online]. Available: www.cleanenergyregulator.gov.au/Infohub/CPM/Unit-shortfall-charges [Accessed 08.04.2020].

AUSTRALIAN GOVERNMENT. 2015b. *About the mechanism* [Online]. Available: www.cleanenergyregulator.gov.au/Infohub/CPM/About-the-mechanism [Accessed 08.04.2020].

AUSTRALIAN GOVERNMENT. 2015c. *INDC submissions. Australia* [Online]. United Nations Framework Convention on Climate Change. Available: www4.unfccc.int/Submissions/INDC/Published%20Documents/Australia/1/Australias%20Intended%20Nationally%20Determined%20Contribution%20to%20a%20new%20Climate%20Change%20Agreement%20-%20August%202015.pdf [Accessed 15.08.2017].

AUSTRALIAN GOVERNMENT. n.d.-a. *Clean Energy Act 2011* [Online]. Federal Register of Legislation. Available: www.legislation.gov.au/Details/C2011A00131/Download [Accessed 02.04.2020].

AUSTRALIAN GOVERNMENT. n.d.-b. *Managing-the-reef. Threats-to-the-reef. Climate-change. What does this mean for habitats? Coral-reefs* [Online]. Available: www.gbrmpa.gov.au/managing-the-reef/threats-to-the-reef/climate-change/what-does-this-mean-for-habitats/coral-reefs%20) [Accessed 05.09.2017].

AUSTRALIANPOLITICS.COM. 2011. *Carbon tax legislation becomes law* [Online]. Available: https://australianpolitics.com/2011/12/09/carbon-tax-legislation-becomes-law.html [Accessed 09.09.2017].

BAILEY, I., MACGILL, I., PASSEY, R. & COMPSTON, H. 2012. The fall (and rise) of carbon pricing in Australia: A political strategy analysis of the carbon pollution reduction scheme. *Environmental Politics*, 21, 691–711.

BANERJEE, S. 2012. *Update on Australia's carbon pricing mechanism* [Online]. Australian Government. Department of Climate Change and Energy Efficiency, International Carbon Action Partnership. Available: https://icapcarbonaction.com/en/?option=com_attach&task=download&id=14 [Accessed 02.10.2017].

BEESON, M. & MCDONALD, M. 2013. The politics of climate change in Australia. *Australian Journal of Politics and History*, 59, 331–348.

CARBON MARKET WATCH. 2015a. *Towards a global carbon market: Prospects for linking the EU ETS to other carbon markets* [Online]. Available: https://carbonmarketwatch.org/wp-content/uploads/2015/05/NC-Towards-a-global-carbon-market-report_web.pdf [Accessed 02.04.2020].

CEMENT INDUSTRY FEDERATION. 2013. *Australia–EU registry linking* [Online]. Available: www.cement.org.au/Portals/0/Documents/Submissions/CIF%20Submission%20on%20Registry%20Options%20to%20Facilitate%20Linking%20of%20Emissions%20Trading%20Schemes.pdf [Accessed 03.04.2020].

COMMONWEALTH OF AUSTRALIA. 2017. *Australia's 7th National communication on climate change* [Online]. Available: https://unfccc.int/files/national_reports/national_communications_and_biennial_reports/application/pdf/024851_australia-nc7-br3-1-aus_natcom_7_br_3_final.pdf [Accessed 02.05.2020].

CROWLEY, K. 2013. Pricing carbon: The politics of climate policy in Australia. *Interdisciplinary Reviews: Climate Change*, 4, 603–613.

CSIRO & THE BUREAU OF METEOROLOGY. n.d. Available: www.climatechangeinaustralia.gov.au/en/ [Accessed 03.05.2018].EU COM. 2012. *Australia and European Commission agree on pathway towards fully linking emissions trading systems* [Online]. Available: https://ec.europa.eu/clima/news/articles/news_2012082801_en [Accessed 19.09.2016].

EU COM. 2013. *Linking EU ETS with Australia: Commission recommends opening formal negotiations* [Online]. European Commission. Available: http://ec.europa.eu/clima/news/articles/news_2013012401_en.htm [Accessed 13.08.2014].

EU COM. n.d. a. *International carbon market* [Online]. Available: https://ec.europa.eu/clima/policies/ets/markets_en [Accessed 20.03.2020].

EU LEX. 2015. *Judgment of the Court (Grand Chamber) of 16 July 2015* [Online]. Available: https://eur-lex.europa.eu/legal-content/EN/TXT/?uri=CELEX%3A62013CJ0425 [Accessed 23.03.2020].

GARNAUT, R. 2008. *The Garnaut climate change review* [Online]. Available: www.hume.vic.gov.au/files/46a4d08c-9a31-4c6d-bed0-9e1c00c093b3/Garnaut_Climate_Change_Review_Report.pdf [Accessed 20.09.2016].

HERALD SUN. 2010. *Kevin Rudd delays plans for emissions trading scheme until 2013* [Online]. Available: www.heraldsun.com.au/news/kevin-rudd-delays-plans-for-emissions-trading-scheme-until-2013/news-story/999d4ea4533e0dc67392fcd6abe587a9 [Accessed 13.11.2017].

HUDSON, M. 2015. *Tax or trade, the war on carbon pricing has been raging for decades* [Online]. Available: https://theconversation.com/tax-or-trade-the-war-on-carbon-pricing-has-been-raging-for-decades-46008 [Accessed 05.09.2018].

ICTSD 2011. Australia, New Zealand plan to link emissions trading schemes. *Bridges*, 15.

IRIGOYEN, C. 2017. *The carbon tax in Australia* [Online]. Available: www.centreforpublicimpact.org/case-study/carbon-tax-australia/ [Accessed].

JOTZO, F. 2012. Australia's carbon price. *Nature Climate Change*, 2, 475.

JOTZO, F. & BETZ, R. 2009. Australia's emissions trading scheme: Opportunities and obstacles for linking. *Climate Policy*, 9, 402–414.

LEWIS, S. & JACOB, P. 2013. *Carbon tax hurts, say ailing firms* [Online]. news.com.au. News Corp Australia Network. Available: www.news.com.au/tablet/carbon-tax-hurts-say-ailing-firms/news-story/0d5d45be8206b4e63d8fb78515116bf9 [Accessed 24.03.2020].

MINCHIN, L. 2006. *Howard blows hot and cold on emissions* [Online]. THE AGE. Available: www.theage.com.au/national/howard-blows-hot-and-cold-on-emissions-20061115-ge3kkq.html [Accessed 13.11.2017].

MÜLLER, P. & SLOMINSKI, P. 2016. Theorizing third country agency in EU rule transfer: Linking the EU Emission Trading System with Norway, Switzerland and Australia. *Journal of European Public Policy*, 23, 814–832.

PALMER, C. 2012. *Carbon price shift to tie Australian govt to European Policy* [Online]. The Conversation. Available: https://theconversation.com/carbon-price-shift-to-tie-australian-govt-to-european-policy-9147 [Accessed 24.03.2020].

PARLIAMENT OF AUSTRALIA. 2000. *The heat is on: Australia's greenhouse future. Emissions trading.* [Online]. Available: www.aph.gov.au/Parliamentary_Business/Committees/Senate/Environment_and_Communications/Completed_inquiries/1999-02/gobalwarm/report/c09 [Accessed 05.09.2018].

PEEL, J. 2014. The Australian carbon pricing mechanism: Promise and pitfalls on the pathway to a clean energy future. *Journal of Law Science and Technology Issue*, 15, 429.

QUIGGIN, J., ADAMSON, D. & QUIGGIN, D. 2014. *Carbon pricing: Early experience and future prospects*, Cheltenham: Edward Elgar Publishing.

ROURKE, A. 2012. *Australian and EU carbon markets to be linked* [Online]. Sydney: The Guardian. Available: www.theguardian.com/environment/2012/aug/28/australia-eu-carbon-markets [Accessed 24.03.2020].

SABIN CENTER FOR CLIMATE CHANGE LAW & COLUMBIA UNIVERSITY. 2015. *European Commission vs. Council for the European Union* [Online]. Available: https://climatecasechart.com/non-us-case/european-commission-v-council-for-the-european-union/ [Accessed 02.04.2019].

SANTIKARN, M., LI, L., THEUER, S. L. H. & HAUG, C. 2018. *A guide to linking emissions trading systems* [Online]. Berlin: Report published by ICAP. Available: www.icapcarbonaction.org/publications [Accessed 03.09.2019].

SPENCER, T. A., SÉNIT, C.-A. & DRUTSCHININ, A. 2012. The Political Economy of Australia's Climate Change and Clean Energy Legislation: Lessons Learned [Online]. SSRN. Available: https://papers.ssrn.com/sol3/papers.cfm?abstract_id=2198316 [Accessed 02.05.2020].

SWOBODA, K., TOMARAS, J. & TALBERG, A. 2012. *Clean energy amendment (International emissions trading and other measures) bill 2012* [Online]. Available: www.aph.gov.au/Parliamentary_Business/Bills_Legislation/bd/bd1213a/13bd037 [Accessed 01.03.2020].

TALBERG, A. & SWOBODA, K. 2013. *Emissions trading schemes around the world* [Online]. Available: https://parlinfo.aph.gov.au/parlInfo/download/library/prspub/2501441/upload_binary/2501441.pdf;fileType=application/pdf [Accessed 08.04.2020].

TAYLOR, L. 2014. *Australia kills off carbon tax* [Online]. Available: www.theguardian.com/environment/2014/jul/17/australia-kills-off-carbon-tax [Accessed 24.03.2020].

TENNANT, M. 2013. *Australia's Carbon Tax Contributing to Record Business Failures* [Online]. The New American. Available: www.thenewamerican.com/world-news/australia/item/14874-australia-s-carbon-tax-contributing-to-record-business-failures [Accessed 24.03.2020].

THE CLIMATE INSTITUTE. 2013. *Emissions reductions from carbon laws cap equivalent to 15 per cent target* [Online]. Sydney: Media Brief 21 November 2013 Available: www.climateinstitute.org.au/verve/_resources/ TCI_MediaBrief_EmissionReductions_21November2013.pdf   [Accessed 13.04.2016].

WHITE, A. 2013. *Why Tony Abbott wants to abolish the carbon price* [Online]. Available: www.theguardian.com/environment/southern-crossroads/2013/ sep/18/tony-abbott-abolish-carbon-price [Accessed 24.03.2020].

WWF AUSTRALIA. 2012. *WWF-Australia submission to the House of Representatives Standing Committee on economics' inquiry into the amendment (international emissions trading and other measures) bill 2012 and associated bills.* [Online]. Available: www.aphref.aph.gov.au_house_committee_economics_cleanenergy2012_submissions_sub08wwfaustralia.pdf   [Accessed 24.03.2020].

WWF AUSTRALIA. 2013. *Australia's emissions trading scheme implications of an early move to a flexible price* [Online]. WWF Policy Brief July 2013. Available: www.wwf.org.au/ArticleDocuments/pub-policy-brief-australiasemissions-trading-scheme-16jul13.pdf [Accessed 24.03.2020].

WWF EU & VAN DEN PLAS, S. 2012. *Australia to join EU ETS by 2018: Europe must put its house in order before then* [Online]. Available: https://wwf. panda.org/wwf_news/?206048/Australia-to-join-EU-ETS-by-2018-Europemust-put-its-house-in-order-before-then [Accessed 24.03.2020].

7 Linking is dominated by domestic political interests, domestic structures, and international developments

The starting point of this research is the EU ETS. Launched in 2005, this instrument has existed for over 15 years. Since then, the EU ETS has passed through a significant process of maturation, during which its initial design and rules were reformed. During the past decade, it has also been somewhat of a hotspot for cooperation and linking activities. The cases analyzed in the previous chapters are those in which linking became rather well-developed, even if not always realized, after being considered for several years. The three linking activities took place between 2008 and 2019 (EU ETS–CAL C&T: 2008–2011; EU ETS–SW ETS: 2008–2019; EU ETS–AUS CPM: 2011–2014). The starting dates of these cases are all relatively close together, falling into the second and third operation phase of the EU ETS. Almost in parallel, the EU ETS reformation process was also initiated. During the period examined for this book, the EU's partners were still in their initial ETS implementing phases and working out the specifications of their ETS designs.

The cases exhibit both common and divergent starting conditions. All four ETSs were based in industrialized and relatively rich jurisdictions with relatively strong economic and knowledge capacities for coping with climate change; however, they differ in exposures to environmental problems and climate change, as well as their emission profiles. The EU and Australia have a high-emitting energy sector, both involving heavily coal-dependent regions (European Environment Agency, 2016, Commonwealth of Australia, 2017). In contrast, in Switzerland, fossil fuel-based energy production plays a smaller role because hydroelectric and nuclear power are predominant, and in California it is the transport sector that accounts for the largest share of GHGs (CARB, 2019c, BAFU, 2017a). Also, all four jurisdictions have different climate policy approaches. In California and Switzerland, the ETS is not the main climate policy tool. In the EU and Australia, the ETS stood at the center of climate policy as the main tool for reducing GHGs. Yet, all four jurisdictions faced some political struggle when implementing the ETS. Interest groups from the industrial or energy sectors expressed opposition to the implementation of each ETS and fought over specific ETS design aspects, for example, the auctioning of allowances. In Australia, where climate policy was strongly contested, the political struggles around the ETS were particularly intense.

The first case, the EU ETS–CAL C&T linking activity, can be seen as a failed case. Linking talks were abandoned at the earliest stage, even before official negotiations could start. The different jurisdictional statuses of the EU and California made this case structurally very complex and politically challenging. But additionally, policymakers did not expect sufficient benefits, e.g. economic or environmental, from the link. Both perceived themselves as climate policy leaders and adapting their domestic ETS policy to the partner's demand did not fit this plan. If California and the EU had ignored these difficulties and linked in spite of the US federal government's reluctant stance on climate change, such an agreement would have sent a strong and symbolic political signal for international climate policy cooperation and commitment to ETS as a policy tool. Perhaps given the fact that in 2018, the EU and California renewed their commitment to intensifying the cooperation between their ETSs, this academic inquiry is especially warranted.

The second case, the EU ETS–SW ETS linking activity, can be seen as a successful case because it led to a ratified linking treaty and the operation of a linked market. It is not so surprising that this has been the most feasible link, as the two ETSs were very similarly designed and Switzerland is geographically, economically, and politically very close to Europe. However, though labelled as an 'easy endeavor', completion of the linking of the two ETSs took over a decade. Also in this case, political-structure factors and conflicting interests, emerging for example in the struggle to include aviation into the linking agreement or, independent of ETS policy, the diplomatic affront perceived after the Swiss vote on migration, significantly prolonged the linking process.

Finally, in the EU ETS–AUS CPM case, the partners came to an informal agreement on a unilateral link starting in 2015, but in the end the link was not implemented. This case received most excitement among policy-makers and other stakeholders, but was also the one confronted with most political conflicts. The success of the case was heavily dependent on external conditions, such as the public's perception of climate change. Here, the most decisive political development, the Australian general elections, initially pushed negotiations to proceed quickly in order to anchor the AUS CPM within the political system.

But ultimately, their outcome, the government turnover, and subsequent abolition of the AUS CPM, caused the link to fail.

In the second and the third observed cases, linking talks were initiated by the EU's partners, Switzerland and Australia. In the Californian case, efforts to start linking were pursued both by EU and Californian policy-makers. It can be established that Switzerland and Australia had stronger motivations for linking than did the EU. For them, the outcome was or would have been an international treaty. Conversely, with California, a link could only be realized in the form of a MoU.

General observations

The comparative perspective on the three observed cases helps to answer the question of why linking, in spite of the high academic and political expectations outlined in the introduction to this book, lacks in progress. The very broad conclusion that can be drawn is that political aspects, such as domestic interests, domestic structural conditions, and international developments, have challenged linking attempts so far. This comes in addition to the technical and economic complexity of linking. It is notable that of the three observed cases, linking was only successful in the one case where strong favorable domestic interests existed and where domestic structures and international developments did not impede linking. This case study analysis suggests that a*t least one of the linking partners must have strong domestic interests in linking, which must persist over time*, for the process to lead to a linked market. In the most successful of the observed cases, EU ETS–SW ETS, Switzerland had a relatively strong interest in linking over the course of more than ten years; even though structural challenges occurred, they were not so big as to impede linking; and international developments rather favored linking.

Furthermore, *the interplay of political factors—domestic interests, structures, and international developments—is decisive for linking*. This underscores a classic neoliberal assumption that the interdependencies and overlaps between domestic and international structures should be an essential element in the analysis of cooperation (Milner, 1992, Keohane, 1984, Frieden, 1999, Moravcsik, 1997). All factors analyzed in this research showed some correlation between each other. The EU ETS–AUS CPM linking efforts demonstrated that even though a singular event, the general elections in Australia, brought an end to the negotiations, it was the mutual reinforcement of several factors, such as stakeholder opposition and agenda setting, behind this event that made linking fail. *Overall, the interplay of the policy (i.e. the design and objective of the ETS) and the politics of both linking and ETS in general are decisive for linking.*

In general, linking politics are very dynamic. As suggested in the conceptual part of this book, linking does not consist of one singular cost-benefit calculation by the responsible policy-makers, but a series of decisions and adjustments of factors. *The political conditioning parameters are not constant over all linking phases but vary over time. Therefore, the same parameters can have both a pushing and a retaining influence or even stop linking at different phases during the process.* For example, in the case EU ETS–AUS CPM the general elections at some point pushed the linking negotiations as Australian policy-makers saw an opportunity to lock-in the ETS, but then the timing of the elections and the government turnover made the whole endeavor fail.

Lessons learned from the practical cases

The analysis of three linking cases can both complement and challenge assumptions from previous research on linking, and further 'unpack' the effects of some influencing factors that have so far received no or very little attention.

For the EU, linking has been first of all motivated by an interest in reinforcing its status as a leader and role model and spreading the standards of the EU ETS. The EU's interest in playing a leadership role has pushed the EU to link, even when economic benefits were absent or limited to an abstract level. This supports assumptions about structural and directional leadership in international environmental cooperation and those authors who argue that frontrunning raises a normative 'standard' which others voluntarily follow. These three linking case studies suggest that behind the interest of taking a leadership position there are other important motivations rooted in the advantages that can be derived from setting regional standards that other regions subsequently adopt. *Linking for the EU has been not only about signaling international commitment and cooperation, but about a vision of expanding ETS as a global policy tool with the EU ETS at the center.* Indeed, this book's research furthers the suggestion that linking is also an attempt by the EU to influence the design and rules of other emissions trading systems (Rutherford, 2014).

Nevertheless, *leadership aspirations do not unequivocally forward linking.* This research observed that the simultaneous aspirations of the EU ETS and the CAL C&T to be a leader or role model actually had an inhibiting impact on linking. Of course, in reality the situation between the EU ETS and the CAL C&T was far more complex. The two systems were not competing for the same regional influence; California, contrary to the EU, aspired to be a North American leader, while the EU had first an OECD-wide and then a generally global focus. Leadership is, as Keohane and Victor (2016) suggest, also about showing a 'convincing' model. It is motivated by the expectation of reputational benefits or losses and by 'being perceived as leader by others' (Keohane and Victor, 2016). This was perceived in the case studies, when policy-makers saw the other ETS as an 'attractive or unattractive' model or linking partner. This was a weak point of the EU ETS, which was perceived as unattractive to other actors, when design flaws led to an oversupply of allowances and cyber frauds let the system appear weak. *The perceived lack of being a convincing model resulted in the EU ETS's loss of some jurisdictions, such as California, as followers.*

Some of these motivations for leadership are closely linked to another finding of this research: *The pursuit of powerful negotiation positions and imbalanced power constellations plays a complex role in linking, going beyond the need for regulative control over the ETS alone.* Being a role model that other jurisdictions follow symbolizes power. But in the more successful of the observed linking cases, Swiss and Australian policy-makers did not necessarily cede to the EU ETS standards because the EU ETS was an attractive model (which, as noted above, was then not realistically the case). To this point, classic arguments from international relations that see power as the capability to pressure international counterparts (Putnam, 1988) and argue that power imbalances can be conducive to cooperation (Milner, 1992) are shown to be appropriate. Swiss and Australian policy-makers accepted EU standards because they depended on the economic and other benefits the link promised. This *interdependency*, and the EU's much stronger influence on the linked market through the EU ETS' size, left the EU's partners in a weaker negotiation position. Even though stakeholders and some policy-makers in Australia and Switzerland maintained concerns about this imbalance in power, these were ultimately unsuccessful in stopping the linking activity. However, no such strong interests in the expected benefits existed in California, and the CAL C&T's functioning did not depend on access to the EU ETS.

In the linking literature, much weight has been given to policy-makers concerns about losing control over the domestic ETS (Flachsland et al., 2009a, Carbon Market Watch, 2015a, Victor, 2015, Comendant and Taschini, 2014, Green et al., 2014a, Ranson and Stavins, 2016). *This research suggests that concerns over the loss of control will succumb to the expectation of greater benefits from the prospect of cooperation,* as has above been described for the cases of Switzerland and Australia.

Another finding of the analysis of these linking cases is that the very core argument for linking, *the promise of an increase in overall economic efficiency, pushes linking first and foremost when utilized on an abstract level, as rhetoric to gain support from stakeholders.* The EU embraced linking even though none of the cases promised concrete economic benefits for the EU. The majority of literature on linking cites very general notion of cooperation taking place in order to achieve competitive advantages (Keohane and Victor, 2016) and lowering costs through an expanding market (Santikarn et al., 2018, Burtraw et al., 2013, Flachsland et al., 2009a, Haites and Mullins, 2001, Ellis and Tirpak, 2006, Tuerk et al., 2009, Jaffe et al., 2009). These benefits might have created a generally positive environment to initiate a linking process but were likely insufficient for the EU to make progress in linking. Yet, the argument was used throughout the EU's official communications, press releases and figures, presumably in order to gain stakeholders' support. This finding complements suspicions raised by Flachsland et al. (2009a), who argued that linking had a value in facilitating the acceptance of climate policy among domestic businesses and the public. However, even though the promise of lower costs after linking might help gain the support of some stakeholders, when such economic benefits do not help very much when they remain too vague. For example, in the EU ETS–CAL C&T case, potential economic gains apparently stayed too abstract to be seen as real advantages. *Thus, this research does not contradict the economic rationale for linking, but it finds that in practice, the value of this argument lies first of all in providing a positive narrative for linking.*

As the Californian example suggests, *in order to be relevant for linking, economic interests must be relatively concrete.* The promise of a better functioning market as a result of linking to the larger EU ETS worked well for the smallest partner, the SW ETS, because it faced liquidity challenges. In the case with Australia, there was an expectation that a cheaper EU ETS allowance price would lower the anticipated much higher emissions abatement costs and thus the burden for the participating Australian companies. *The expectations of the carbon price and its effects had very complex and nonlinear impacts on linking.* The role of the carbon price was different for each case. For the EU ETS–CAL C&T, forecasts on the carbon price in a linked market were either unspecific or could not be made at all, because the EUAs price fluctuated strongly and the CAL C&T had not started its operation. Thus, rather than a concretely expected allowance price, it was more likely the risk of an 'uncertain target' and a potentially less stringent EU ETS that made Californian policy-makers hesitant to link. California's explicit deviation from the EU model when introducing a minimum allowance price into the CAL C&T hints at the fact that Californian policy-makers perceived the EUAs price as too low. In Switzerland, even though the cheaper EUAs price was calculated as a benefit for Swiss industry for a long time, the then-significant price increase after 2018 did not change Swiss policy-makers' positive attitude on linking and did not stop the ratification of the treaty. In Australia, the cheaper EUAs price was widely seen as beneficial, even though this fact hinted at the weak stringency of the EU ETS. All in all, these observations suggest that *the carbon price of a partner's system, and whether it is seen as disadvantageous or advantageous, does not necessarily depend on whether the price is high or low, but how it fits to domestic needs.* This assumption is connected to the argument raised by several authors who state that the actual gain from linking depends on the match with the linking partner (Santikarn et al., 2018, Flachsland et al., 2009a, Haites, 2014, Ranson and Stavins, 2016) and the domestic circumstances (Gulbrandsen et al., 2019). Thus, even though a high carbon price is generally seen as better for the environment or climate, because it incentivizes companies to act more strongly to implement low carbon technologies, what matters politically is less a question of price than what policy-makers consider desirable for the economy and the ETS constituents.

Central for the international climate policy debate is that the research carried out for this book gives reason to assume that *environmental and climate policy objectives are secondary to other interests and motivations in linking.* In the examined cases where linking progressed, it was not primarily realized in order to gain climate benefits. This finding puts in doubt the idea that the best solution to reduce emissions worldwide would be a global carbon market, and the general notion that ETS linking would enhance climate policy (Santikarn et al., 2018, Edenhofer et al., 2007, Bodansky, 2002). Rather, it reinforces studies that are more skeptical about the environmental or climate benefits of linking (Green, 2017, Gulbrandsen et al., 2019). Climate and environmental benefits appear to have played a minor role in the observed cases. No jurisdiction has (publicly) contemplated the option of increasing climate policy objectives after linking. Even more important is that, except maybe for California, none of the EU's partners cited a lack of environmental effectiveness in the EU ETS, due to its low carbon prices, as a reason to stop pursuing linking. Policy-makers proceeded with linking in spite of the environmental stringency-related concerns raised by Australian and Swiss stakeholders. These notions stand in contrast to the fact that ETS is first of all a climate policy tool. This contradiction can be explained by at least four ideas: First, it is technically and politically very difficult to assess the partner system's real environmental stringency and ambition and how it will develop over time, because too many (uncertain) conditions and political developments have to be taken into account. Second, the climate policy benefits that can be expected from linking, such as improved cooperation, might be too abstract; it is almost impossible to forecast what climate benefit will result from a link. Third, linking shifts benefits from the local to the global or international level, as it is more difficult to estimate where emissions reductions and other environmental benefits occur in a linked market. Thus, these benefits may be less 'tangible' to stakeholders and increase their concerns. Fourth, how much climate and environmental aspects are prioritized during the linking activity probably also depends on how capable stakeholders are to voice advocacy or opposition and push for their (environment- and climate-related) concerns.

This book's analysis highlights the relevance of domestic concerns for linking. The fear of domestic disadvantages, ranging from increasing consumer costs to the general uncertainty of where emissions reductions would occur after linking, are important to stakeholder groups in all EU partner jurisdictions. Only few authors consider stakeholders' general role in linking (e.g. Ranson and Stavins [2013] or Santikarn et al. [2018]), and even fewer scholars examine specific stakeholders activities or/and the role of specific constituents (for example, Jevnaker and Wettestad [2014]). This book takes a first step toward unpacking the effects of stakeholder involvement on linking. However, this research also concludes that more in-depth investigation is needed in order to gain a clearer picture about how stakeholder activity influences policy-makers' decisions to link or not.

In all four jurisdictions, policy-makers named the motivation of satisfying industry's interest as a driver behind the initiation of linking discussions. Yet, in the cases of EU ETS–SW ETS and EU ETS–AUS CPM, industrial stakeholder groups' support was heterogeneous. While some sectors such as aviation in Switzerland and power generation in Australia were more negative toward ETS and/or linking, large industrial companies active on the international markets viewed linking positively. Environmental groups showed concerns about linking in all three cases, but it is not clear to what degree their concerns influenced linking progress and outcomes. *A careful assumption could be made that industrial stakeholders obtained more weight in the linking activities in Switzerland and Australia, as linking was fostered despite environmental groups' concerns.* Such a finding could be underscored by arguments made by international relations scholars who have suggested that industrial groups often have more ability to pressure policy-makers as they

can threaten with closing production plants and removing job opportunities (Offe, 1972). For example, in ETS politics, industrial groups
threaten the risk of 'carbon leakage', referring to the movement of
industrial production to regions with lower or no climate policy regulation, resulting in the 'leakage' of GHG emissions to other parts of
the world.

*A remarkable aspect, when looking at these three cases, is that mean-
ingful stakeholder activity was directly concerned less with linking than
with developments in the broader political setting.* In fact, the societal stakeholder activity that effectively impacted linking was mostly
concerned with aspects other than linking, or those more indirectly
connected to linking. For example, in Australia, large parts of the public
opposed the overall introduction of carbon pricing and national climate
policy and were concerned with the general elections. In Switzerland,
the discontent with the broader imbalanced relationship to the EU, and
the strong EU influence on Swiss politics through, for example, being
the most important trading partner, were expressed in the Swiss referendum on migration; this brought negotiations to a temporary stop.
In California, the general design and establishment of climate legislation and local priorities were fought out between environmental groups,
parts of the industrial sector and policy-makers.

Also in the EU, most groups, including both industrial and societal stakeholders, were concerned with the functioning of the EU ETS.
However, the research carried out for this book found very little activity
from EU industrial or societal stakeholders during each of the three
linking attempts. This research suggests that, w*hile stakeholders—
ranging from environmental and industrial groups to the public—took an
interest in the respective ETS implementation processes in all jurisdictions,
linking was too specific a topic for many.* As a representative of an
industrial interest group stated, industrial stakeholders are generally
in favor of linking and support a global carbon market, but this perspective becomes more complicated whenever linking becomes concrete
(Interview 5).

*Another explanation for the limited stakeholder opposition or advo-
cacy could be that industrial groups, societal groups, and the general
public were not well informed about linking.* On the one hand, ETS
linking is a technically complex issue for which a deep knowledge of the
instrument is required. For example, a representative of an industrial
interest group argued that many of its constituents were not informed
enough to deal with the issue of linking (Interview 5). Additionally, in
the case EU ETS–SW ETS, a lack of transparency from the EU COM
had been criticized (Carbon Market Watch, 2015b), as the linking talks

and negotiations were often held behind closed doors, with little information being released to stakeholders (Interviews 2, 6). Similarly, in California, public opinion polls found that even five years into the existence of the CAL C&T, in 2017, 56% of the interviewed Californians had never heard of the ETS (Baldassare et al., 2017). One further reason for the lack of EU interest groups' activity could be that, as constituents of the larger linking partner, they did not expect large impacts on the private sector or societal and environmental conditions. Also, EU environmental groups might have perceived the higher allowance prices of the EU's linking partner systems as proof of their stronger environmental ambitions and thus expected that linking with these partners would actually improve environmental objectives also within the EU.

Last but not least, an explanation for the lack of interest groups' opposition and advocacy could be that this research has not completely penetrated the stakeholder dynamics within the observed jurisdictions. For example, in the EU, future research could differentiate stakeholder activity in and between the 27 member states, as well as their relationship to the EU organs.

The research carried out for this book found an outstanding role for *responsible authorities and government officials, who influenced linking beyond their mere negotiation task*. For example, personnel entrepreneurship by administrative or political personnel fostered linking in both Switzerland and Australia. Furthermore, the examined cases largely confirm Wettestad and Jevnaker's (2014) assertion that the role of the EU COM was generally a pushing one. Yet, the EU COM played various other roles related to linking. For example, some interviewees' perception of closed negotiations and a lack of transparency from the EU COM (Interviews 2, 16) hints at somewhat of an informational asymmetry leaning in favor of the EU COM. Information availability and—exclusivity—can bring with it advantages and powerful positioning (Moravcsik, 1997). The questions of whether and how potential political power imbalances between the responsible authorities, such as between CARB and the EU COM, impacted the negotiations requires further investigation, with one interviewee suggesting that the EU COM did not treat all partner authorities equally (Interview 6). An additional topic in need of further research is the role of transnational actors, such as the World Bank and ICAP, which in this research were observed to have facilitated learning and exchange for linking, however, the specific interests behind their activities remain somewhat opaque.

Additionally, broader political developments unrelated to ETS have led to more failure than success in the observed linking activities. Structural developments external to ETS can be relevant obstacles to linking. In the Australian case, the political momentum of general elections and the subsequent political agenda derailed the link to the EU ETS. But also in Switzerland, challenging aspects appeared when the public referendum on migration raised questions about the general relationship between the EU and Switzerland and temporarily halted the linking process. *Politics external to the linking initiatives, for example migration policies, had a strong influence on linking.* This is supported by interviewees who stated that ETSs in general are susceptible to developments and aspects that are not directly ETS-related (Interview 16). For the Australian case, where negotiations moved fast in spite of technical differences, it can be assumed that *external political developments superseded technical challenges.* These findings underscore the necessity to pay special attention to the ETS setup process and its systemic context. In this regard, future linking studies could draw more intensively on knowledge from those international relations authors who examine how the legislative procedures, the decision-making rules and law (Risse-Kappen, 1991), or how parliamentary or congressional majorities may affect the chances of domestic adoption of a policy (Putnam, 1988).

This research also claims that the factor which has been named as the single most important predictor for linking by some scholars, geographical proximity (Ranson and Stavins, 2013), actually has a much more complex effect. For the EU ETS–AUS CPM case, the absence of proximity, and the resulting state of less trading and competition, was observed as positive. Then, in the case of EU ETS–SW ETS, the Swiss' public questioning of their political and historical relationship to their proximate EU neighbors brought negotiations to a temporary halt. Negotiations between the EU and Switzerland took over a decade in spite of their geographical closeness and the similar design of their ETSs. At the same time, a much less tangible aspect of proximity, the existence of trust, built through close coordination, knowledge, and ultimately confidence in the partner jurisdiction appeared to have fostered linking. Thus, it can be concluded that *whether political, cultural, and geographical relationships influence linking positively or negatively depends on the interplay with other conditions for linking.*

The international setting of linking, how partners perceive each other, and the international ETS and climate policy landscape are essential ingredients in linking politics. The analyzed linking processes demonstrate the need for more careful examination of the assumption that the more ETSs emerge globally, the more linking is likely to progress. *The expansion of the global carbon market does not automatically drive linking but has more complex effects.* On the one hand, global ETS expansion was a good argument for EU policy-makers to counter arguments that

ETS and climate policy regulations would lead to domestic competitive disadvantages. Also, the establishment of market mechanisms under the Kyoto Protocol and the Paris Agreement create a positive atmosphere for the initiation of linking. On the other hand, the emergence of 'other actors', or new ETSs on the international stage, has had a different effect in the concretely examined cases. Here, *new ETSs also increased the competition for linking partners, and this can divert a responsible agency's attention from one jurisdiction to another.* For instance, the launch of an ETS in Québec provided a linking opportunity to California that was much more attractive than the EU ETS, hindering the possible EU ETS–CAL C&T link. Moreover, Australia's decision *against* linking to another ETS, New Zealand ETS, created a positive environment for its own linking negotiations with the EU, because EU negotiators saw it as symbol for strong commitment to their mutual linking endeavor.

All in all, this book's initial assumption that *politics play an indispensable role in the practice of linking appears to be substantiated. Linking is not only a coordination problem, in the sense that technicalities have to be coordinated, but a cooperation problem, in which domestic interests, domestic structures, and international influences all have to be taken into account.* All three cases experienced very different domestic circumstances and faced challenges at different stages. Still, a common finding for all three is that the main obstacles have mainly been political. Political factors often exist behind technical differences in ETS systems or arise during negotiations. One possible explanation is that each ETS is designed to satisfy domestic needs and demands (Gulbrandsen et al., 2019, Green et al., 2014b). Thus, the ETS's design reflects the government's decisions, as constrained by domestic preferences, as well as local political and administrative realities (Green et al., 2014b, Gulbrandsen et al., 2019). As an ETS represents the political priorities of a region, or its 'political comfort zones' (Interview 2), linking becomes a sensitive topic. When an ETS is established, a domestic consensus is reached that represents what was politically feasible in the jurisdiction in question. Linking, with all its implications, means opening up and, in some cases, even renegotiating this consensus. In those cases where linking is considered at the same time as the ETS implementation process, policy-makers may fear the additional broadening of this consensus building through the partner jurisdiction's demands.

Last but not least, *this book underscores the administrative and political,* or in other words, *the governance complexity of linking* (Interview 3). On both the domestic and international levels, the stakes are very high, as is the number of actors involved, ranging from societal groups to different governance levels. Much learning and expertise is required to deal with the technical complexity of linking. As some interviewees remarked, the regulative effort of linking is underestimated (Interview 8). Even though this research's findings on the different roles of stakeholders are not extensive, they underscore the need for future analyses to take into account a more differentiated view on the many actors involved in linking.

These case analyses support the notion that linking, as form of cooperation, is not only an outcome but also a multilevel and repeated-game process. It is a step-by-step process, in which gathering new information and learning is essential and decisions are constantly adapted to developments in the changing domestic and international structures. This book's findings can be related to studies that emphasize on the learning process between ETSs and which argue that decisions over political 'realities', that is domestic circumstances, have impeded convergence of ETSs (Gulbrandsen et al., 2019, Wettestad and Gulbrandsen, 2013, Bang et al., 2017).

Considerations for the EU

As discussed in the introduction to this work, some authors (Ranson and Stavins, 2012, Jaffe et al., 2009, Comendant and Taschini, 2014, Bodansky et al., 2016) have argued that linking could be (part of) the solution to the global lack of climate cooperation and provide a bottom-up architecture for international climate policy. But this analysis has shown that linking has a different starting point. Rather than as the solution to a conflict, policy-makers often see it as 'a step on top' of national climate action, or as a second step to optimize the ETS following its introduction. This distinction is significant for the general theoretical assumptions, because it suggests that when linking fails, the associated losses or the lack of cooperation may not be as strongly resented, compared to if linking were seen as more central to conflict resolution. Turning this argument around, one could argue that often, *few concrete or tangible losses are connected to 'not linking'.* Thus, potentially higher gains must be expected to justify the administrative effort needed for linking. This assumption was confirmed by certain experts who expressed the view that in general, linking bears few concrete advantages. They argued that often, the expected gains from linking do not outweigh the efforts and costs (Interviews 3, 16). When considered alongside the claim that linking is not necessarily beneficial for environment or climate change mitigation objectives, the element of insufficient gains may lead to a far-reaching conclusion as pointed out by some scholars (Green, 2017, Gulbrandsen et al., 2019): *Linking is not under all circumstances good for the common cause of making quick—but sustainable—progress in climate change mitigation.*

For the EU and the European ETS policy, linking and the expansion of the EU ETS model represented the opportunity to gain (or sustain) a leadership position in the international climate policy landscape. However, the observed cases have shown that for now *linking is unlikely to function as a vehicle for EU leadership.* Not only has the EU ETS faced many technical challenges during the past decade, EU linking activities have also proven to be rather complicated and long-term endeavors. Even in the case of EU ETS–SW ETS, which was expected to be a simple process due to the EU's relationship with Switzerland and the similarity in their ETS designs, linking took more than a decade. The analysis of the EU's linking activities has revealed that the EU ETS and broader ETS politics are highly dependent on domestic interests, structures, and developments; such factors range from the complex decision-making process and the difficulty to integrate and consult all necessary stakeholders, to climate policy-independent developments such as general elections. Also, how the EU COM can deal with subnational actors in international relations, when the main negotiating partner is not a sovereign nation state, should not be ignored in the field of international climate policy cooperation, as this research demonstrated in the case of California. The examined cases have shown that under such circumstances, challenges clearly go beyond strict legal status-related aspects, but that also here, politics and diplomatic behavior are essential.

Future EU COM attempts to link will have to be accommodated into the EU's climate policy strategy and be treated as part of the EU's internal policy process on general directions of climate policy. As of the time of writing, the EU's climate policy targets for 2030 are established as 'domestic targets', which in principle excludes their achievement through external credits and transfers (such as linking). The EU's emissions reductions targets are currently been revised and it remains to be seen, in how far the transfer of foreign or international allowances will play a role in the future climate plan.

References

BAFU. 2017a. *Switzerland's greenhouse gas inventory 1990–2015. National inventory report. Including reporting elements under the Kyoto Protocol* [Online]. Available: www.infras.ch/media/filer_public/7b/7c/7b7c5db3-6426-45db-88e2-a9ca3bb88a95/che-2017-apr-nir.pdf [Accessed 02.05.2020].

BALDASSARE, M., BONNER, D., KORDUS, D. & LOPES, L. 2017. *Californians & the environment* [Online]. Available: www.ppic.org/wp-content/uploads/s_717mbs.pdf [Accessed 24.10.2017].

BANG, G., VICTOR, D. G. & ANDRESEN, S. 2017. California's Cap-and-Trade system: Diffusion and lessons. *Global Environmental Politics*, 17, 12–30.

BODANSKY, D. 2002. *US climate policy after Kyoto: Elements for success* [Online]. Washington, DC: Carnegie Endowment for International Peace. Available: https://carnegieendowment.org/2002/03/25/u.s.-climate-policy-after-kyoto-elements-for-success-pub-937 [Accessed 14.06.2018].

BODANSKY, D. M., HOEDL, S. A., METCALF, G. E. & STAVINS, R. N. 2016. Facilitating linkage of climate policies through the Paris outcome. *Climate Policy*, 16, 956–972.

BURTRAW, D., PALMER, K. L., MUNNINGS, C., WEBER, P. & WOERMAN, M. 2013. Linking by degrees: Incremental alignment of cap-and-trade markets [Online]. Resources for the Future 13-04. Available: www.rff.org/publications/working-papers/linking-by-degrees-incremental-alignment-of-cap-and-trade-markets/ [Accessed 12.03.2017].

CARB. 2019c. *GHG current California emission inventory data. 2019 GHG inventory* [Online]. Available: https://ww2.arb.ca.gov/ghg-inventory-data [Accessed 02.05.2020].

CARBON MARKET WATCH. 2015a. *Towards a global carbon market. Prospects for linking the EU ETS to other carbon markets* [Online]. Available: https://carbonmarketwatch.org/wp-content/uploads/2015/05/NC-Towards-a-global-carbon-market-report_web.pdf [Accessed 02.04.2020].

CARBON MARKET WATCH. 2015b. *Towards a global carbon market. Risks of linking the EU ETS to other carbon markets* [Online]. Available: https://carbonmarketwatch.org/wp-content/uploads/2015/05/NC-Towards-a-global-carbon-market-PB_web.pdf [Accessed 04.04.2020].

COMENDANT, C. & TASCHINI, L. 2014. *Submission to the inquiry by the House of Commons Select Committee on Energy and Climate Change on 'Linking Emissions Trading Systems'* [Online]. Centre for Climate Change Economics and Policy Grantham Research Institute on Climate Change and the Environment. Available: www.cccep.ac.uk/Publications/Policy/docs/Comendant-and-Taschini-policy-paper-April-2014.pdf [Accessed 20.07.2016].

COMMONWEALTH OF AUSTRALIA. 2017. *Australia's 7th National communication on climate change* [Online]. Available: https://unfccc.int/files/national_reports/national_communications_and_biennial_reports/application/pdf/024851_australia-nc7-br3-1-aus_natcom_7_br_3_final.pdf [Accessed 02.05.2020].

EDENHOFER, O., FLACHSLAND, C. & MARSCHINSKI, R. 2007. *Towards a global CO_2 market* [Online]. Available: https://pdfs.semanticscholar.org/29fd/f66e2341bc63d2d57b9b431a42ddf2a328b5.pdf [Accessed 03.02.2020].

ELLIS, J. & TIRPAK, D. 2006. Linking GHG Emission Trading Systems and Markets [Online]. Organisation for Economic Co-operation Development,

International Energy Agency. Available: www.oecd.org/environment/cc/37672298.pdf [Accessed 03.03.2019].

EUROPEAN ENVIRONMENT AGENCY. 2016. *Sectoral greenhouse gas emissions by IPCC sector* [Online]. Available: www.eea.europa.eu/data-and-maps/daviz/change-of-co2-eq-emissions-2#tab-dashboard-01 [Accessed 15.7.2018].

FLACHSLAND, C., MARSCHINSKI, R. & EDENHOFER, O. 2009a. To link or not to link: Benefits and disadvantages of linking cap-and-trade systems. *Climate Policy*, 9, 358–372.

FRIEDEN, J. A. 1999. *Actors and preferences in the international relations*, Princeton, Princeton University Press.

GREEN, J. F. 2017. Don't link carbon markets. *Nature News*, 543, 484.

GREEN, J. F., STERNER, T. & WAGNER, G. 2014a. A balance of bottom-up and top-down in linking climate policies. *Nature Climate Change*, 4, 1064–1067.

GREEN, J. F., STERNER, T. & WAGNER, G. 2014b. *The politics of market linkage: Linking Domestic climate policies with international political economy* [Online]. Fondazione Eni Enrico Mattei Nota di Lavoro 64.2014. Available: www.econstor.eu/bitstream/10419/102003/1/NDL2014-064.pdf [Accessed 03.04.2018].

GULBRANDSEN, L. H., WETTESTAD, J., VICTOR, D. G. & UNDERDAL, A. 2019. The political roots of divergence in carbon market design: implications for linking. *Climate Policy*, 19, 427–438.

HAITES, E. 2014. *Lessons learned from linking emissions trading systems: General principles and applications* [Online]. Washington, DC: Partnership for Market Readiness (PMR). Available: www.thepmr.org/system/files/documents/PMR%20Technical%20Note%207.pdf [Accessed 09.09.2019].

HAITES, E. & MULLINS, F. 2001. Linking domestic and industry greenhouse gas emission trading systems [Online]. Margaree Consultants. Available: http://citeseerx.ist.psu.edu/viewdoc/download?doi=10.1.1.512.9323&rep=repl&type=pdf [Accessed 12.12.2018].

JAFFE, J., RANSON, M. & STAVINS, R. N. 2009. Linking tradable permit systems: A key element of emerging international climate policy architecture. *Ecology LQ*, 36, 789.

WETTESTAD, J. & JEVNAKER, T. 2014. The EU's quest for linked carbon markets. *In:* CHERRY, T. L., HOVI, J. & MCEVOY, D. M. (eds.) *Toward a new climate agreement: Conflict, resolution and governance.* London: Routledge.

KEOHANE, R. O. 1984. *After hegemony: Cooperation and discord in the world political economy*, Princeton: Princeton University Press.

KEOHANE, R. O. & VICTOR, D. G. 2016. Cooperation and discord in global climate policy. *Nature Climate Change*, 6, 570.

MILNER, H. 1992. International theories of cooperation among nations. *World Politics*, 44, 466–496.

MORAVCSIK, A. 1997. Taking preferences seriously: A liberal theory of international politics. *International Organization*, 51, 513–553.

OFFE, C. 1972. *Klassenstrukturen: Zur Analyse spätkapitalistischer Gesellschaftssysteme*, Frankfurt: Fischer Taschenbuch Verlag.

PUTNAM, R. D. 1988. Diplomacy and domestic politics: The logic of two-level games. *International Organization*, 42, 427–460.

RANSON, M. & STAVINS, R. N. 2012. *Post-Durban climate policy architecture based on linkage of Cap-and-Trade systems* [Online]. National Bureau of Economic Research. Available: www.belfercenter.org/sites/default/files/files/publication/ranson-stavins_dp51.pdf [Accessed 13.03.2020].

RANSON, M. & STAVINS, R. N. 2013. *Linkage of greenhouse gas emissions trading systems: Learning from experience* [Online]. Washington: Resources for the Future RFF DP13-42. Available: www.rff.org/publications/working-papers/linkage-of-greenhouse-gas-emissions-trading-systems-learning-from-experience/ [Accessed 03.09.2017].

RANSON, M. & STAVINS, R. N. 2016. Linkage of greenhouse gas emissions trading systems: Learning from experience. *Climate Policy*, 16, 284–300.

RISSE-KAPPEN, T. 1991. Public opinion, domestic structure, and foreign policy in liberal democracies. *World Politics*, 43, 479–512.

RUTHERFORD, A. P. 2014. Linking emissions trading schemes: Lessons from the EU-Swiss ETSs. *Carbon & Climate Law Review*, 8, 282–290.

SANTIKARN, M., LI, L., THEUER, S. L. H. & HAUG, C. 2018. *A guide to linking emissions trading systems* [Online]. Berlin: Report published by ICAP. Available: www.icapcarbonaction.org/publications [Accessed 03.09.2019].

TUERK, A., MEHLING, M., FLACHSLAND, C. & STERK, W. 2009. Linking carbon markets: Concepts, case studies and pathways. *Climate Policy*, 9, 341–357.

VICTOR, D. G. 2015. *The case for climate clubs* [Online]. International Centre for Trade and Sustainable Development (ICTSD). Available: www.e15initiative.org/ [Accessed 27.07.2018].

WETTESTAD, J. & GULBRANDSEN, L. H. 2013. The evolution of carbon trading systems: Waves, design and diffusion. *International Cooperation*, 40.

8 Linking in the climate policy debate and future prospects

ETS linking cannot be seen independently of the developments in international climate policy and politics. *In a number of ways, linking faces similar cooperation problems as those found in the international climate policy process under the UNFCCC*, where domestic circumstances and the anticipation of (non) action often contributes to the slow pace of negotiations. Linking is characterized by difficulties in coordinating policies and making domestic interests compatible with global challenges. All in all, one can also interpret the slow progress of linking, as well as its myriad difficulties, as a cause for pessimism. It shows that even jurisdictions with similar climate policy approaches and, in the observed cases, similar ambitions, struggle to cooperate beyond mere 'talk and exchange'. In this sense, a general observation of international climate policy is also applicable to linking: all constituents can agree that cooperation is important, but the more concrete the level of cooperation required gets, the more difficult agreeing on action becomes.

Overall, the three linking cases of the EU with California, Switzerland, and Australia as partners can be seen as products of their time. On the one hand, their initiation reflects a strong optimism about market policies that existed directly following the Kyoto Protocol's 2005 entry into force. On the other hand, the subsequent slow pace with which linking progressed and the failure of some linking attempts contributed to policy-makers' growing understanding of the technical and political complexity of ETS as an instrument. Possibly, the slow progress of linking also reflects the lesson that market mechanisms cannot be seen as the one 'silver bullet' to cope with the climate challenge, as has already been suggested by Jevnaker and Wettestad (2016). The cases examined here illustrate demonstrate that most ETSs were designed to suit domestic objectives and circumstances, and that such domestic approaches cannot easily be transformed into international cooperation.

Linking politics can be seen as attempts at bilateral cooperation that operate in addition or complement to the UNFCCC process. Against the backdrop of insufficient national pledges to achieve the Paris Agreement's target of keeping the global temperature under 2°C (UN Environment, 2018), scholars have discussed how alternative forms of cooperation could provide additional mitigation opportunities. Following the initially stated logic that ETS linking could, in theory, raise climate policy ambition, one may assume that linking could be a vehicle for additional mitigation (e.g. if jurisdictions were to enhance their mitigation commitments through the prospect of lowered costs after linking). However, the case studies analyzed for this book suggest the need for a more skeptical perspective on this idea. If carried through, linking is indeed *a special form of cooperation between jurisdictions that envisions both more and deeper connections to 'like-minded' partners.* However, as noted above, there is no evidence yet that linking actually drives jurisdictions to more ambitious climate policy. Furthermore, the slow progress of linking worldwide, coupled with the complexity of political and policy dynamics, leads to the conclusion that linking is not generating additional emissions reductions, at least at this moment in time. It is also currently difficult to prove whether the prospect of linking has effectively incentivized any jurisdictions to implement new ETSs.

The observed cases give reason to believe that linking is not a form of cooperation that will promote climate policy through increased commitments or increased emissions reductions. However, it could be an opportunity to showcase the attractiveness of a certain climate policy tool and the learning process that takes place during linking efforts, regardless of whether they end with a linking agreement or not, has an added value for climate policy as a whole.[1] Even though this book suggests that, to date, linking has not acted as means to strengthen climate political ambition (at least not directly), such argumentation does not necessarily mean that climate policy cannot benefit from linking. Apart from the symbolic value of a strong climate policy commitment, the learning and interjurisdictional exchange that arises in a linking, even if no linking treaty results from it, facilitates mutual understanding and expertise, and may even improve the jurisdictions' conceptions of climate and ETS policy in the long term. In this sense, ETS linking efforts, such as the observed cases, might help policy-makers prepare for the (future) implementation of national commitments under the UNFCCC, given linking's contribution to enhancing cooperation, exchange, and technical learning.

Future prospects for more progress in linking?

This book is based on the assumption that *linking remains relevant and could be put on the agenda of policy-makers in the mid- or long-term future* even though at the time of writing linking has shown little progress. This prospective on linking was affirmed unanimously by the interviewees for this research. A logical follow-up question to this work's findings would be: *'Under what circumstances could linking experience more progress in the future?'* This outlook chapter attempts to provide possible answers. While many of the examined barriers would remain relevant should linking efforts resume or increase in the near future, linking research has discussed several ideas that could potentially change the current difficult linking dynamics. These possibilities involve either changes made to the linking process itself, for example, through limiting the extent of the links; or changes in systemic conditions, such as the availability of new partner ETSs worldwide (Interviews 2, 10, 16), an increase in abatement costs (Interviews 12, 13, 15), or planned activities under Article 6 of the Paris Agreement (Interviews 12, 14).

Restricted linking

In light of the difficulties associated with bilateral full linking of ETSs, various scholars have argued for *restricted linking* as a potentially more viable option (Lazarus et al., 2015). The rationale for a restricted or limited form of linking is that it would provide policy-makers and jurisdictions with more time to overcome policy differences (Quemin and de Perthuis, 2019) and allow them to keep stronger control over their own markets while still being able to benefit from some of the advantages that come with linking. Restricted linking can be used as interim measure and a testing ground for a fuller form of linking. For example, Burtraw et al. (2013) propose a stepwise linking or *'linking by degrees'* to break down the complex linking process in several steps and over time to gradually amass most of the elements of a fully linked market. This includes a gradual learning process that gives participants the chance to build confidence and gain knowledge of other systems. In doing so, it can help them to overcome obstacles (Quemin and de Perthuis, 2019, Burtraw et al., 2013).

The most common way to implement a restricted link is the above-discussed *one-way linking,* or unilateral acceptance of emissions allowances. Examples include links to the CDM, the initial connection of the Norway ETS to the EU ETS, and the intended first phase of linking between the AUS CPM and the EU ETS. One-way links could be administered under the new market mechanisms of the Paris Agreement's Article 6.

Other options to keep more control over the ETS are to introduce a fixed *quota* of allowed exchanges, to define an *exchange* or *discount rate*, or to put a *fee on foreign credits*. Quotas are typically set at a fraction of the total compliance obligations, and they limit the overall sum of credits transferred. They can ensure that certain amounts of emissions reductions are achieved domestically and that the level of allowance flow is kept within a politically acceptable range (Lazarus et al., 2015). Even though quotas per se are attractive as they do not change the overall emissions abatement levels, it can be politically difficult to set a quota at an effective level (Lazarus et al., 2015), and uncertainty underlies the price formation process (Quemin and de Perthuis, 2019). Another option to restrict linking would be the setting of exchange rates, which adjust the value of the linking partner's allowances, equating a certain amount to one ton of domestic compliance obligation (Lazarus et al., 2015). However, exchange rates have complex consequences and can affect the overall cost efficiency and total abatement within the linked jurisdictions (Lazarus et al., 2015). Last but not least, jurisdictions could introduce a *fee* or *tax* on foreign allowances. Taxes can help to manage better distributional outcomes and price fluctuations, however, also here it is difficult to find an effective price level; a too high rate, for example, would make the link ineffective (Quemin and de Perthuis, 2019).

These options for restricted linking have an appeal, because they enable a try-out link and allow for some of the benefits such as increased cost effectiveness and market liquidity to be realized. However, these approaches only diminish or postpone the need for the technical alignment of ETS design elements (Lazarus et al., 2015). Also, they may still cause disputes over what is politically acceptable, such as the level of a quota or exchange rate. This is due to the fact that many of the political conditions, such as power constellations or concerns over financial flows, would still be in place. For example, experts interviewed for this research have acknowledged that allowance export rates would be difficult to implement because it would be politically challenging to find an adequate level for the rate and then justify it to domestic stakeholders (Interviews 2, 14). Although these options have been discussed by academics and practitioners, at the time of writing they have not been applied with much vigor, and they are mostly limited to the quota-based links used to offset schemes such as the CDM and likely the future activity under Article 6 of the Paris Agreement. Thus, it can cautiously be concluded that the value of restricted linking can be seen in it enabling a pathway and playing field for linked ETSs, but that for now this strategy does not seem to trigger short-term increases in bilateral linkages as a whole.

In addition to restricted linking, one option to improve and potentially streamline the linking process is the installation of an overseeing regional or international institution. Some authors have contemplated an institution, such as an international clearing house (Edenhofer et al., 2007) or a central carbon bank (Green, 2017), to help with the coordination between domestic carbon pricing approaches. Yet, as some of the authors themselves admit (Green, 2017), these ideas seem unlikely to bear fruit in today's political landscape that is characterized by struggles to agree on centralized climate policies. Thus, this book does not further consider such proposals at this point.

Expansion of the global market, emergence of alternative linking partners: China

As observed in the introduction to this book, worldwide carbon market initiatives are proliferating. The simple logic is that the more ETSs exist worldwide, the more alternative linking options are available. Even though in the previous chapter this book had suggested that more ETSs do not directly lead to more links, the emergence of one very large actor has started to influence the ETS landscape significantly: China's development of pilot ETSs and its plan to launch a national ETS have gained much attention.

Already in 2009, the then responsible national authority, the National Development Reform Commission (NDRC), announced its aspirations to build a national ETS in China by 2015; it started pilot systems in several regions (Ochs and Haibing, 2010), namely Beijing, Chongqing, Guangdong, Hubei, Shanghai, Shenzhen, and Tianjin (ICAP, 2015a). In September 2015, before the adoption of the Paris Agreement, the Chinese president announced that China's national carbon market would be launched in 2017 (Volcovici, 2015, The White House President Barack Obama, 2015). The plan was proclaimed by some as the 'the start of a new era in climate policy for the country' (Swartz, 2016), but, as of the date of writing, only a political launch was realized in December 2017. In 2019, the newly formed Ministry of Ecology and Environment, responsible for the ETS, was still working on setting national ETS regulations, completing the market infrastructure, and putting in place systems for reporting and verification (ICAP, 2020a). Once fully operational, the ETS is expected to regulate only the power sector, with a potential gradual expansion to other sectors (Duan et al., 2018). The approximately 1,700 participating companies would cover more than three billion tons of CO_2e or about 30% of national emissions (ICAP, 2019a, Duan et al., 2018). Currently, the seven mentioned provincial ETSs, plus an additional one in Fujian, are already operating (ICAP, 2020a).

In spite of these uncertainties, linking to the Chinese ETS has been on the EU's discussion desk for several years (Interviews 5, 11, 15, 16) (Reklev, 2016). Huge EU investments in ETS capacity-building measures for China were motivated by the hope that once China implemented a national ETS, this system would have fewer design incompatibilities with the EU ETS due to this close cooperation (Interview 15). Yet, these activities are being carried out with the knowledge that such a link is still far from realization (Interviews 11, 15), and to date no linking negotiations have been suggested. Conversely, Chinese political representatives have remained more conservative on the prospects for linking. At an event of the UNFCCC COP in 2017, a representative of the Chinese national government concretely stated that no link was being planned, because China's national ETS would have to first start operations and function properly before any linking activities could be considered.[2]

In principle, a link between a Chinese national ETS and the EU ETS would have major symbolic meaning for the international climate policy arena and could alter global views on linking (Zeng et al., 2018) (Interviews 11, 16). Furthermore, concrete climate policy collaboration between two of the world's largest economic blocs and top GHG emitters could have tremendous political consequences (Interview 11). Such a union might have geopolitical meaning, signaling a very close connection between the EU and China to other countries, such as the United States. Once technically negotiated, it would also establish the specific design and rules for the world's largest ETS shared market and, depending on the linking regulations agreed upon, it could give impetus to the application of the EU ETS model worldwide.

Overall, many of the political obstacles found in this research's cases studies also play a role here. A link between the EU ETS and a Chinese national ETS would alter the carbon price and the financial distribution balance within the EU ETS and beyond. It is likely that EU-based companies would become net buyers of Chinese allowances, which to date have been significantly cheaper than EUAs. Between 2019 and 2020, prices of the Chinese pilot ETSs fluctuated between 0.3 EUR and 11.3 EUR, compared to the 17.4 EUR to 29.4 EUR price of EUAs (ICAP, 2020d). It is quite complex to predict whether or not a price convergence after linking and thus strong market alteration would be politically desirable. Even though, in principle, the option of cheap allowances could appeal to some industrial stakeholders in the EU, the resulting flow of funds to China would be met with criticism (Interview 5). Furthermore, it might be expected that following the EU COM's long struggle for reform and a carbon price increase, there is currently little appetite for low EUA prices. On China's end, keeping allowances cheap to ensure an incoming flow of funds could appeal to some stakeholders (Interview 17). Such stakeholders might see linking to the EU ETS as an opportunity for a second boom in selling cheap emissions reduction certificates, following what happened when China became the main seller of CERs generated under the CDM (Interview 17).

A link between the EU ETS and a Chinese national ETS could also raise concerns about environmental integrity in light of some of the Chinese ETS's technical design elements. For example, some experts have highlighted that the Chinese ETS approach will likely lead to electricity carbon leakage (Zeng et al., 2018). Furthermore, a link to a national Chinese ETS would change the power constellations for the EU. The much larger Chinese market would have an outsized influence over the EU ETS. For the first time, the EU would be in a position, where another jurisdiction wields greater influence over linking negotiations and the linked market. Ceding control to Chinese decision makers is seen very critically by the EU COM (Interview 15) and also by EU industrial stakeholders (Interview 5). Experts suggest there is somewhat weak trust between EU and Chinese policy-makers. Not only are EU policy-makers somewhat unconvinced of the credibility of the Chinese ETS, Chinese policy-makers perceive cooperation processes with the EU as very complicated and time-consuming (Interview 15, 17).

In China, carbon market linking is a politically sensitive issue. Chinese policy-makers responsible for ETS may fear a reputational loss, material consequences (such as job loss or even severe legal reprimands) if they are blamed for an unwanted crash in carbon prices (Interview 17). Overall, the vast difference between China and the EU's political systems, as well as their climate politics and policies, represents a major hurdle to linking their ETSs (Interview 17).

This very brief look into potential motivations and challenges for a link between the EU and China suggests that despite the EU interest in gaining China as a partner, a link between both ETSs is unlikely to happen in the near future. ETS cooperation would be technically complex and politically challenging. Linking challenges would be exacerbated by structural aspects, such as the different political systems in China and the EU. Thus, it is unlikely that an EU–China link will add to the number of global linkages in the near-term future.

Changing conditions: Rising abatement costs and Article 6 of the Paris Agreement

The introduction to this book emphasizes that the Paris Agreement under the UNFCCC includes passages setting a framework for the transfer of emissions reductions achievements between countries. Article 6.1 calls for international cooperation on NDCs. Articles 6.2 and 6.3 frame how international mitigation outcomes can be transferred between parties. Articles 6.4 and 6.7 create a new mechanism for generating emissions reductions in a host country that can be used by another party (Marcu, 2017b). Another option would be that under the Paris Agreement's updating mechanism, countries decide to enhance their emissions reduction targets. For example, a more ambitious EU GHG reduction target, would have an impact on the EU ETS (Koch et al., 2016, Marcu et al., 2016). Indirectly this could also influence linking prospects, for example, if the allowance prices in an ETS increased and the jurisdictions decided to seek cheaper abatement opportunities.

Overall, about half of the parties of the Paris Agreement plan to use market mechanisms to meet their national commitments (Howard, 2018). Several authors agree that Article 6 creates a positive international setting for carbon market linking by establishing an institutional framework for it (Howard, 2018, Marcu, 2017a, Mehling et al., 2018). Principally, Article 6 could further motivate countries to pursue linking to other ETSs. It could generate demand and supply for allowances to be traded in a unilateral link, such as occurred with the CDM.

Nonetheless, at the time of completing this book, it is still not clear what the exact rules for carbon markets will look like. Important elaboration and clarification between parties is still required before countries can start implementing Article 6 (Jevnaker and Wettestad, 2016). Experts have highlighted risks attached to the international transfer of emissions reductions, e.g. with respect to the environmental integrity of market bases policy instruments and the risk of countries selling reductions that have effectively not occurred (or so-called 'hot air') (La Hoz Theuer et al., 2019, Schneider and La Hoz Theuer, 2019, Schneider et al., 2018). At the center of discussion stand the accounting rules under the Paris Agreement and the risk of double counting emissions reductions; this could occur if emissions reductions are accounted under national commitments but also at the same time sold to other parties (Müller and Michaelowa, 2019, La Hoz Theuer et al., 2019, Schneider et al., 2019). With growing frustration over the difficulties associated with designing effective rules for Article 6, skeptical voices seem more justified. In 2016, Jevnaker and Wettestad stated that the political challenges related to ETS linking had not disappeared with the adoption of the Paris Agreement. The same is true today. The Paris Agreement has not changed countries' focus on domestic politics and benefits (Interview 8). Findings of this study that suggest the importance of domestic interests and structures would also be relevant for prospective links or transfers under the Paris Agreement.

Finally, this book will briefly consider another prospect. Several experts have argued that interest in linking could be renewed in the event of rising abatements costs and heightened mitigation objectives (Interviews 3, 7, 12, 13, 15). In such a situation, stakeholders can be expected to demand the equalizing of economic conditions that impact competition and, as such, policy-makers will face more pressure to find cheaper abatement opportunities (Interviews 8, 12). Even though this book cannot pursue a full analysis of how such a situation would influence any of the reviewed cases, it can point toward a broader tendency. Action on climate change, both for adaptation and mitigation, is expected to become more expensive in the future (McKinsey&Company, 2009). Explanations include that, at some point, the 'low hanging fruits', i.e. cheap and easily feasible mitigation technologies, will all be achieved, and the remaining emissions reduction options will be more expensive. Furthermore, the severity of climate change impacts and the risks of reaching climate tipping points will increase and thus strengthen the time pressure on policy-makers. The mere expectation of increasing abatement costs can put more pressure on policy-makers to find cost-efficient, or 'industry-friendly', solutions and to implement well-developed climate policy instruments, which by then will include ETSs. As linking is expected to lower the overall costs of ETSs, policy-makers may well turn to the option of linking when abatement costs rise. Ultimately, raising abatement costs would intervene in the cost-benefit calculations made for linking, possibly swinging the balance in favor of linking and a large carbon market. However, even with the increase in expected economic benefits from linking, the political challenges noted in the examined case studies will also have some weight in this future cost-benefit analysis.

This book was born out of a desire to understand the discord between the political and academic interests in linking, as well as the lack of widespread adoption of linking. This research's examination of specific cases provides an explanation for this situation based on the existence of broader political factors that influence national agendas and various actors' interests. This chapter's short outlook has shown that even though some ideas exist on how this situation could change in the future—through restricted linking, an EU link to a national Chinese ETS, or as a means of fulfilling Article 6 of the Paris Agreement—none of these options are likely to significantly increase bilateral linkages in the short-term future. Ultimately, while they do strengthen the enabling conditions for linking, they still fail to eradicate the underlying political challenges. However, time—and increasing costs for mitigating climate change—may at some point prove to be the game changers for linking.

Notes

1 Also for the successful link between the CAL C&T and the QUE C&T that is not analyzed here, interviewees have suggested that California's ability to demonstrate a well-working ETS model served as a main driver (Interviews 8, 14, 19).
2 The author participated in the event 'Exploring the potential for carbon clubs and linked carbon markets in the Asia-Pacific Region' held on 13.11.2017 at the Korean Pavilion/COP 23 in Bonn, Germany, and took note of this statement.

References

BURTRAW, D., PALMER, K. L., MUNNINGS, C., WEBER, P. & WOERMAN, M. 2013. *Linking by degrees: Incremental alignment of Cap-and-Trade markets* [Online]. Resources for the Future 13-04. Available: www.rff.org/publications/working-papers/linking-by-degrees-incremental-alignment-of-cap-and-trade-markets/ [Accessed 12.03.2017].

DUAN, M., QI, S. & WU, L. 2018. Designing China's national carbon emissions trading system in a transitional period. *Climate Policy*, 18, 1–6.

EDENHOFER, O., FLACHSLAND, C. & MARSCHINSKI, R. 2007. *Towards a global CO_2 market* [Online]. Available: https://pdfs.semanticscholar.org/29fd/f66e2341bc63d2d57b9b431a42ddf2a328b5.pdf [Accessed 03.02.2020].

GREEN, J. F. 2017. Don't link carbon markets. *Nature News*, 543, 484.

HOWARD, A. 2018. *Accounting for bottom-up carbon trading under the Paris agreement* [Online]. C2ES Center for Climate and Energy Solutions. Available: www.c2es.org/site/assets/uploads/2018/04/accounting-bottom-up-carbon-trading-paris-agreement.pdf [Accessed 23.01.2020].

ICAP. 2015a. *Emissions trading worldwide: International carbon action partnership status report 2014* [Online]. Available: https://icapcarbonaction.com/en/?option=com_attach&task=download&id=349 [Accessed 13.08.2017].

ICAP. 2019a. *Emissions trading worldwide international carbon action partnership (ICAP) status report 2018* [Online]. Berlin. Available: https://icapcarbonaction.com/en/?option=com_attach&task=download&id=625 [Accessed 23.05.2019].

ICAP. 2020a. *Emissions trading worldwide: International carbon action Partnership (ICAP) status report 2019* [Online]. Berlin. Available: https://icapcarbonaction.com/en/?option=com_attach&task=download&id=677 [Accessed 02.05.2020].

ICAP. 2020d. *Allowance price explorer* [Online]. Available: https://icapcarbonaction.com/en/ets-prices [Accessed 05.05.2020].

JEVNAKER, T. & WETTESTAD, J. 2016. Linked carbon markets: Silver bullet, or castle in the air? *Climate Law*, 6, 142–151.

KOCH, N., GROSJEAN, G., FUSS, S. & EDENHOFER, O. 2016. Politics matters: Regulatory events as catalysts for price formation under cap-and-trade. *Journal of Environmental Economics and Management*, 78, 121–139.

LA HOZ THEUER, S., SCHNEIDER, L. & BROEKHOFF, D. 2019. When less is more: Limits to international transfers under Article 6 of the Paris Agreement. *Climate Policy*, 19, 401–413.

LAZARUS, M., SCHNEIDER, L., LEE, C. & ASSELT, H. V. 2015. *Options and issues for restricted linking of emissions trading systems* [Online]. ICAP Policy Paper. Available: https://icapcarbonaction.com/en/?option=com_attach&task=download&id=279 [Accessed 02.05.2020].

MARCU, A. 2017a. *Article 6 of the Paris agreement: Reflections on party submissions before Marrakech* [Online]. International Centre for Trade Sustainable Development Background Paper. Available: https://secureservercdn.net/160.153.137.163/z7r.689.myftpupload.com/wp-content/uploads/2018/12/article_6_of_the_paris_agreement_ii_final_0.pdf [Accessed 02.05.2020].

MARCU, A. 2017b. *Governance of Article 6 of the Paris agreement and lessons learned from the Kyoto Protocol* [Online]. Centre for International Governance Innovation. Fixing Climate Governance Series. Paper No. 4 — May 2017 Available: www.cigionline.org/sites/default/files/documents/Fixing%20Climate%20Governance%20Paper%20no.4%20WEB.pdf [Accessed 02.05.2020].

MARCU, A., ALBEROLA, E., CANEILL, J.-Y., MAZZONI, M., SCHLEICHER, S., STOEFS, W., VAILLES, C. & VANGENECHTEN, D. 2016. *2018 state of the EU ETS report* [Online]. CEPS Brussels. Available: www.ictsd.org/sites/default/files/20180416_2018_state_of_eu_ets_report_-_final_all_logos_.pdf [Accessed 02.05.2020].

MCKINSEY&COMPANY. 2009. *Pathways to a low-carbon economy. Version 2 of the global greenhouse gas abatement cost curve* [Online]. Available: www.mckinsey.com/~/media/McKinsey/Business%20Functions/Sustainability/Our%20Insights/Pathways%20to%20a%20low%20carbon%20economy/Pathways%20to%20a%20low%20carbon%20economy.ashx [Accessed 02.05.2020].

MEHLING, M. A., METCALF, G. E. & STAVINS, R. N. 2018. Linking climate policies to advance global mitigation. *Science*, 359, 997–998.

MÜLLER, B. & MICHAELOWA, A. 2019. How to operationalize accounting under Article 6 market mechanisms of the Paris Agreement. *Climate Policy*, 19, 812–819.

OCHS, A. & HAIBING, M. 2010. *China may Cap-and-Trade before US* [Online]. China dialogue. Available: www.chinadialogue.net/article/show/single/en/3804-China-may-cap-and-trade-before-US [Accessed 02.04.2020].

QUEMIN, S. & DE PERTHUIS, C. 2019. Transitional restricted linkage between emissions trading schemes. *Environmental and Resource Economics*, *74*, 1–32.

REKLEV, S. 2016. *UK envoy talks up EU-China carbon market link-media* [Online]. Carbon Pulse. Available: https://carbon-pulse.com/14705/ [Accessed 20.04.2018].

SCHNEIDER, L., DUAN, M., STAVINS, R., KIZZIER, K., BROEKHOFF, D., JOTZO, F., WINKLER, H., LAZARUS, M., HOWARD, A. & HOOD, C. 2019. Double counting and the Paris Agreement rulebook. *Science*, 366, 180–183.

SCHNEIDER, L. & LA HOZ THEUER, S. 2019. Environmental integrity of international carbon market mechanisms under the Paris Agreement. *Climate Policy*, 19, 386–400.

SCHNEIDER, L., WARNECKE, C., DAY, T. & KACHI, A. 2018. *Operationalizing an 'Overall mitigation in global emissions' under Article 6 of the Paris agreement* [Online]. Berlin: New Climate Institute. Available: https://newclimate.org/wp-content/uploads/2018/11/Operationalising-OMGE-in-Article6.pdf [Accessed 02.05.2020].

SWARTZ, J. 2016. *China's national emissions trading system* [Online]. Geneva: ICTSD Series on Climate Change Architecture. Available: www.ieta.org/resources/China/Chinas_National_ETS_Implications_for_Carbon_Markets_and_Trade_ICTSD_March2016_Jeff_Swartz.pdf [Accessed 02.04.2018].

UN ENVIRONMENT. 2018. *Emissions gap report 2018* [Online]. United Nation Environment Programme. Available: https://wedocs.unep.org/bitstream/handle/20.500.11822/26895/EGR2018_FullReport_EN.pdf?sequence=1&isAllowed=y [Accessed 04.04.2019].

VOLCOVICI, V. 2015. *China to announce 2017 launch of carbon market, officials say* [Online]. Reuters. Available: www.reuters.com/article/us-usa-china-climatechange/china-to-announce-2017-launch-of-carbon-market-officials-say-idUSKCN0RP00W20150925 [Accessed 09.04.2019].

THE WHITE HOUSE PRESIDENT BARACK OBAMA. 2015. *Fact sheet: The United States and China issue Joint Presidential Statement on climate change with new domestic policy commitments and a common vision for an ambitious global climate agreement in Paris* [Online]. Office of the Press Secretary. Available: https://obamawhitehouse.archives.gov/the-press-office/2015/09/25/fact-sheet-united-states-and-china-issue-joint-presidential-statement [Accessed 02.03.2020].

ZENG, Y., WEISHAAR, S. E. & VEDDER, H. H. B. 2018. Electricity regulation in the Chinese national emissions trading scheme (ETS): Lessons for carbon leakage and linkage with the EU ETS. *Climate Policy*, 18, 1246–1259.

Appendix

Interview information

All interviews were realized by the author in the time period of January 2015–August 2018, via telephone, Skype, or in person. Whenever interviewees gave their permission, the interview was recorded and transcribed. Otherwise, minutes of the interview were prepared by the author. The datasets generated during the research process for this book are not publicly accessible but are available from the corresponding author on reasonable request.

Alphabetical list of interviewed institutions

- Adelphi research
- Adelphi research
- Australian national governmental authority
- Australian National University (ANU)
- Bundesamt für Umwelt Schweiz (BAFU)
- Bundesamt für Umwelt Schweiz (BAFU)
- Bund der Industrie (BDI)
- Bundesministeriums für Umwelt, Naturschutz und nukleare Sicherheit (BMU)
- California Air Resources Board
- California Air Resources Board
- Chinese expert from EU based public policy consultancy
- EU Commission DG Climate
- EU Commission DG Energy
- EU Parliament
- International Carbon Action Partnership (ICAP)
- Mercator Research Institute on Global Commons and Climate Change (MCC)

- New York University
- Swedish Environmental Research Institute (IVL)
- Umweltbundesamt Germany (UBA)

List of interview numbers and dates as referred to in the text

Interview 1 (27.01.2015, Berlin), Interview 2 (30.06.2015, Berlin), Interview 3 (25.08.2015, Berlin), Interview 4 (23.09.2015, Berlin), Interview 5 (23.02.2016, Berlin), Interview 6 (06.04.2016, Berlin), Interview 7 (15.04.2016, Skype/telephone), Interview 8 (27.06.2016, Skype/telephone), Interview 9 (07.09.2016, Skype/telephone), Interview 10 (04.11.2016, Skype/telephone), Interview 11 (16.12.2016, Skype/telephone), Interview 12 (13.01.2017, Skype/telephone), Interview 13 (19.01.2017, Skype/telephone), Interview 14 (24.1.2017, Skype/telephone), Interview 15 (11.04.2017, Skype/telephone), Interview 16 (17.05.2017, Berlin), Interview 17 (09.01.2018, Berlin), Interview 18 (04.05.2018, Skype/telephone), Interview 19 (03.08.2018, Skype/telephone)

Index

agenda setting 9, 23, 29–30, 116; EU ETS–AUS CPM 87, 90, 99, 107; EU ETS–CAL C&T 58–59; EU ETS–SW ETS 73

allocation of allowances 18, 20; EU 40; California 53; Switzerland 67; EU ETS–AUS CPM 91

ambition of climate policy and ETS 3, 7, 18, 25, 104, 106, 114–115; Australia 85, 87

Assembly Bill 32 Global Warming Solutions Act' (AB 32) 51

auction see auctioning

auctioning 1, 17–18, 27; Australia 82, 98; California 52–53; EU 42–43; EU ETS–SW ETS 71; Switzerland 67, 69

Australia Carbon Pricing Mechanism design 82–83

Australia climate policy 80–82

Australia climate targets 80, 82, 84

aviation 43, 69, 72–75, 98, 104

back-loading 42

California Cap-and-Trade Program design 52–53

California climate policy 50–53

California climate targets 51

cap 1, 3, 18, 20, 24–26; California 52; EU 40, 43; reduction factor 18; Switzerland 67

carbon price development under linking 3, 17, 19, 24, 102–103, 119; EU ETS–AUS CPM 84–86; ETS–CAL C&T 55; EU ETS–SW ETS 70, 75

Carbon Pricing Leadership Coalition (CPLC) 4

certificates see emissions allowance definition; see also Clean Development Mechanism (CDM); EU; Switzerland; Australia; offsets; EU ETS—AUS CPM; EU ETS—CAL C&T; EU ETS—SW ETS

Certified Emissions Reductions (CER) 15

China national ETS design 118–119

China provincial ETSs 118

Clean Development Mechanism (CDM) 5, 6, 15, 19, 116–117; Australia 85; EU 40, 43, 54; Switzerland 67; see also offsets; EU ETS—AUS CPM; EU ETS—CAL C&T; EU ETS—SW ETS

Clean Energy Act Australia 81

clearing house 118

climate and environmental policy related interests 23, 25–26, 103–106, 121; EU ETS—AUS CPM 85, 89–90; EU ETS—CAL C&T 56, 59, 98; EU ETS—China national ETS 120; EU ETS—SW ETS 70–71, 74; see also environmental integrity; EU ETS—AUS CPM; EU ETS—CAL C&T; EU ETS—China national ETS; EU ETS—SW ETS

climate policy rationale of linking 3–4, 25, 103–104

climate policy targets influencing linking 3, 7, 26, 29, 121; EU ETS—AUS CPM 85; EU ETS—CAL C&T 56; EU ETS—SW ETS 75

CO_2 Act Switzerland 66–67

CO_2 Levy 66–67, 69, 71

co-benefits 23, 26–27, 56, 71, 85–86

competitive distortions 3, 22, 24–25, 102, 108, 122; EU ETS—AUS CPM 85, 89, 91

compliance periods 18, 20, 40

corporate sector *see* private sector; EU ETS—AUS CPM; EU ETS—CAL C&T; EU ETS—China national ETS; EU ETS implementation; EU ETS—SW ETS

cost containment measures 19–20; *see also* price floor; price ceiling

credits *see* emissions allowance definition 1; *see also* Clean Development Mechanism (CDM); Australia; EU; Switzerland; *see also* offsets; EU ETS—AUS CPM; EU ETS—CAL C&T; EU ETS—SW ETS

cyber attacks 18, 42, 59, 101

distribution of financial gains and losses 7, 18, 24, 89, 91, 117, 119

economic efficiency *see* economic interests in linking; EU; EU ETS—EU AUS CPM; EU ETS—CAL C&T; EU ETS—SW ETS

economic interests in linking 3, 7, 9, 22–24, 98, 100, 122; EU 102; EU ETS—EU AUS CPM 81, 84–85, 91–92, 101; EU ETS—CAL C&T 55–56, 59, 102; EU ETS—SW ETS 69

economic rationale of linking *see* economic interests in linking; EU; EU ETS—EU AUS CPM; EU ETS—CAL C&T; EU ETS—SW ETS

effectiveness of ETS 6, 18, 25, 40, 42, 103; EU ETS—AUS CPM 89–90;

EU ETS—CAL C&T 56–57; EU ETS— SW ETS 67

emissions allowance definition 1

enforcement 18, 20

entrepreneurship 44, 74, 90, 106

environmental costs of linking 4, 26, 103–104, 109

environmental interests *see* climate and environmental policy related interests; EU ETS—AUS CPM; EU ETS—CAL C&T; EU ETS—China national ETS; EU ETS—SW ETS; *see also* environmental integrity; EU ETS—AUS CPM; EU ETS—CAL C&T; EU ETS—China national ETS; EU ETS—SW ETS

environmental groups and stakeholders 31, 40; EU ETS—AUS CPM 85, 90, 104; EU ETS—CAL C&T 51, 59, 104–106; EU ETS—SW ETS 74, 89, 104

environmental integrity 18, 23, 25–26, 121; EU ETS—AUS CPM 85–86, 89; EU ETS—CAL C&T 56, 59; EU ETS—China national ETS 120; EU ETS—SW ETS 74

environmental justice movement California 51–52; see also environmental groups and stakeholders; EU ETS—AUS CPM; EU ETS—CAL C&T; EU ETS—SW ETS

ETS design elements overview 17–20

ETS size as conditioning factor for linking 9, 23, 28–29, 101; EU ETS—AUS CPM 86–87; EU ETS—CAL C&T 57; EU ETS—China national ETS 120; EU ETS—SW ETS 72

EU climate policy 39

EU climate targets 39, 42–43, 110

EU ETS Directive 39, 43–45, 68

EU ETS design 42–43

EU ETS implementation 40–42

exchange or discount rate 117

expert interviews 9, 126–127

financial revenue from allowance auctions 23, 27, 43, 53; EU ETS—AUS CPM 83, 85–86; EU ETS—CAL C&T 53, 56; EU ETS—SW ETS 71

flexibility mechanisms 19–20, 40

frontrunner *see* leadership; directional; EU ETS—AUS CPM; EU ETS—CAL C&T; EU ETS—SW ETS

full linking 15, 24, 83, 116

Garnaut Report 83

global carbon market 3, 10, 40, 53, 86, 100–107, 118

government turnover 30, 83, 87–88, 99–100

imbalance of power 57–58, 72, 75, 86–87, 101, 105–106

income from allowance auctions *see* financial revenue from allowance auctions; EU ETS—AUS CPM; EU ETS—CAL C&T; EU ETS—SW ETS

indirect linking 15–16, 25, 31

inductive research 9

industrial groups *see* private sector; EU ETS—AUS CPM; EU ETS—CAL C&T; EU ETS—China national ETS; EU ETS implementation; EU ETS—SW ETS

International Carbon Action Partnership (ICAP) 6, 32, 60, 91, 106

international climate policy negotiations 4, 121, 115

international conditions for linking 23, 32–33, 99, 107–109, 114, 121; EU ETS—AUS CPM 91–92; EU ETS—CAL C&T 60; EU ETS—SW ETS 75

international cooperation theory 21–22

international linking agreement concept 16–17

international systemic conditions *see* international conditions for linking 23, 32–33, 99, 107–109, 114, 121;

EU ETS—AUS CPM 91–92; EU ETS—CAL C&T 60; EU ETS—SW ETS 75

Jurisdiction and legal status 30, 58, 110

Kazakhstan ETS 5

Kyoto Protocol 33, 39, 60, 114; commitments 39, 60, 66, 88, 91; market mechanisms 5, 43, 60, 88, 108; ratification 50, 60, 81

leadership 9, 23, 27–28, 100–101, 110; directional 27, 100; EU ETS—AUS CPM 86, 90; EU ETS—CAL C&T, 57, 60, 98, 100–101; EU ETS—SW ETS 71

learning 3, 32, 106, 108–109, 115–116

legal dispute 55, 59, 88

legislative procedures influencing linking 29–30, 107; EU ETS—AUS CPM 87–88; EU ETS—CAL C&T 58; EU ETS—SW ETS 68, 73–75

linking agreement adoption 16–17; EU ETS—AUS CPM 83; EU ETS—SW ETS 68

linking by degrees 7, 116

linking definition 1, 15

linking process 15–16; EU ETS—AUS CPM 83–84; EU ETS—CAL C&T 53–54; EU ETS—SW ETS 68

long-term 10, 42, 110, 115–116

loss of control over ETS regulation 7, 29, 58, 101

low carbon innovation and technology 23, 27, 42

marginal abatement costs 3, 122 *see also* rising abatements costs

market liquidity 3, 70, 102

market oversight 18, 20, 26

market stability reserve 42, 53

measurement, reporting and verification (MRV) 18, 20, 25

Memorandum of Understanding (MoU) 17, 44, 58, 99

Mexico carbon market or ETS 5

minimum price *see* price floor

Nationally Determined
  Contributions (NDCs) 4–5, 42,
  80, 121
New Zealand 5, 8, 92, 108
non-binding linking arrangement
  see Memorandum of
  Understanding (MoU)
Norway Emission Trading System
  44–45, 116

obstacles linking EU ETS—China
  ETS 119–120
offsets 5–6, 15, 19–20, 25–26, 117;
  EU ETS—AUS CPM 83–84,
  87; EU ETS—CAL C&T 52,
  54; EU ETS—SW ETS 66–67,
  69; *see also* Clean Development
  Mechanism (CDM); Australia;
  EU; Switzerland
overall goal *see* collective goal
over allocation *see* oversupply of
  emissions allowances; EU ETS—
  AUS CPM; EU ETS—CAL C&T;
  EU ETS—SW ETS
oversupply of emissions allowances
  25, 40–41, 101; EU ETS—AUS
  CPM 85–86, 89; EU ETS—CAL
  C&T 56–57, 59; EU ETS—SW
  ETS 70–71, 74; *see also* price
  collapse EU ETS

Paris Agreement 4–6, 9, 33, 108,
  118; Article 6 5, 116–117, 121;
  commitments 39, 42, 66–67, 80, 115
phases of the linking process 16, 54,
  69, 73, 84, 100
political conditions of linking
  framework 20–33
political parties 30–31, 74–75,
  81, 89–90
powerful bargaining position 9, 23,
  27–29, 101, 106, 117; EU ETS—
  AUS CPM 86–87; EU ETS—CAL
  C&T 57–58; EU ETS—China
  national ETS 120; EU ETS—SW
  ETS 72, 75
price ceiling 19–20, 53, 55, 59, 82, 84
price collapse EU ETS 40–41, 55–56,
  59, 70, 85 *see also* oversupply of
  emissions allowances

price floor 19–20, 55, 82, 84, 103
price volatility 30
private sector 24, 31, 98, 105–106,
  39–40, 43; EU ETS—AUS CPM
  81, 86, 88, 90, 104; EU ETS—CAL
  C&T 55, 59; EU ETS– China
  national ETS 120; EU ETS
  implementation; EU ETS—SW
  ETS 66–67, 74–75, 104
proximity cultural, geographic,
  political 7, 27, 31–32; EU ETS—
  AUS CPM 91–92; EU ETS—CAL
  C&T 56; EU ETS—SW ETS
  72–73, 105, 107
public and society 30, 98, 102, 105;
  EU ETS—AUS CPM 80–82, 86,
  88, 90–91, 105; EU ETS—CAL
  C&T 50, 53, 106; EU ETS—SW
  ETS 70, 105, 107
public referendum 51, 68, 73–74,
  105, 107

Québec Cap-and-Trade System 8, 60
quotas for ETS 7, 54, 67, 117

rational choice 9, 21
recycling of revenues 43, 53
Regional Greenhouse Gas Initiative
  (RGGI) 5, 8, 31
registry 18, 42, 69
reputational benefits 27, 57, 101
restricted linking 7, 10, 116–117, 123
rising abatements costs 121–122
role model *see* leadership; EU ETS—
  AUS CPM; EU ETS—CAL C&T;
  EU ETS—SW ETS

scenarios of linked carbon
  markets 6, 17
scope and coverage 20
South Korea ETS 5, 8
stakeholder advocacy 23, 30–31,
  104–106; EU ETS—AUS CPM
  88–89; EU ETS—CAL C&T 56,
  59; EU ETS—SW ETS 67, 74
stakeholder consultation process 74,
  110; *see also* stakeholder advocacy;
  EU ETS—AUS CPM; EU ETS—
  CAL C&T; EU ETS—SW ETS;
  see also stakeholder opposition;

EU ETS—AUS CPM; EU ETS—CAL C&T; EU ETS—SW ETS

stakeholder opposition 7, 23–24, 26, 28, 30–31, 98–99, 104–106; EU ETS—AUS CPM 81–82, 88–90, 105; EU ETS—CAL C&T 51–52, 56, 59; EU ETS—SW ETS 73–75

stringency of ETS 18–19, 23, 25–26; EU ETS—AUS CPM 89, 103; EU ETS—CAL C&T 56, 58; EU ETS—SW ETS 103

structural conditions of linking 22, 29, 99, 106, 120; EU ETS—AUS CPM 87–88; EU ETS—CAL C&T 55; EU ETS—SW ETS 73, 99

subnational actors and action 5, 16, 30, 50, 58–60, 110

Switzerland climate policy 66–67

Switzerland climate target 66–67, 75

Switzerland Emissions Trading System design 66–67

symbolic value of linking 3, 28, 57, 71, 115, 119

targets *see* climate policy targets influencing linking 3, 7, 26, 29, 121; EU ETS—AUS CPM 85; EU ETS—CAL C&T 56; EU ETS—SW ETS 75

technically feasibility of linking 6–8, 17–20, 99, 107–110, 117; EU ETS—AUS CPM 84, 88, 91; EU ETS—CAL C&T 54–56, 59; EU ETS—China national ETS 120; EU ETS—SW ETS 68–69, 73

trust 23, 31–32, 91–92, 107, 120

United Nations Framework Convention on Climate Change (UNFCCC) 4, 9, 60, 80, 91, 114–115, 119–121; *see also* international climate policy negotiations 4, 121, 115

US national emissions trading 53, 57, 60, 91

Western Climate Initiative (WCI) 52

World Bank 5–6, 32, 106

Printed in the United States
by Baker & Taylor Publisher Services